AF344572

S. Vaidya, L.C. Jain and H. Yoshida (Eds.)

Advanced Computational Intelligence Paradigms in Healthcare-2

Studies in Computational Intelligence, Volume 65

Editor-in-chief
Prof. Janusz Kacprzyk
Systems Research Institute
Polish Academy of Sciences
ul. Newelska 6
01-447 Warsaw
Poland
E-mail: kacprzyk@ibspan.waw.pl

S. Vaidya
L.C. Jain
H. Yoshida
(Eds.)

Advanced Computational Intelligence Paradigms in Healthcare-2

With 40 Figures and 11 Tables

Springer

S. Vaidya
Grant Medical Foundation
Ruby Hall Clinic
40, Sassoon Road
411001, Pune
India

H. Yoshida
Harvard Medical School
Massachusetts General Hospital
Department of Radiology
75, Blossom Court
Boston MA 02114
USA

L.C. Jain
University of South Australia
School of Electrical & Info Engineering
Knowledge-Based Intelligent Engineering
Mawson Lakes Campus
Adelaide SA 5095
Australia
E-mail:- Lakhmi.jain@unisa.edu.au

Library of Congress Control Number: 2006934860

ISSN print edition: 1860-949X
ISSN electronic edition: 1860-9503
ISBN 978-3-540-72374-5 Springer Berlin Heidelberg New York

Springer is a part of Springer Science+Business Media
springer.com
© Springer-Verlag Berlin Heidelberg 2007

Cover design: deblik, Berlin
Typesetting by the SPi using a Springer LaTeX macro package
Printed on acid-free paper   SPIN: 12057771   89/SPi   5 4 3 2 1 0

Foreword

This second volume of the book "Advanced Computational Intelligence Paradigms in Healthcare" highlights recent advances in applying computational intelligence to healthcare issues. This book will serve as an interesting and useful resource for health professionals, academics, students, and computer scientists, since it illustrates the current diverse applications of computational intelligence to healthcare practice including topics such as (i) analysing synthetic character technologies such as assessing skills in dealing with trauma patients or obtaining informed consent as well as training medical students in interacting with paediatric patients (ii) menu generation in web-based lifestyle counselling systems, (iii) evaluating models used for studying factors influencing IT acceptance in healthcare practice, (iv) archiving and communicating medical image databases, (v) the use of the electrocardiogram in evaluation and management of patients as well as (vi) rehabilitation and health care for severely disabled people.

Associate Professor Raymond Tedman
School of Medicine
Griffith University
Queensland
Australia

Preface

The goal of healthcare is to maintain or improve human health. To achieve this goal, healthcare systems have evolved considerably over the past years. More and more sophisticated information technologies and intelligent paradigms have been employed in the healthcare systems for delivering effective healthcare to the patients. The computers have made it possible to easily access and process a large amount of information at a relatively low cost and high speed. Computational intelligence is becoming one of the key technologies for healthcare systems to evolve further, because intelligent paradigms such as artificial neural networks, multiagent systems, and genetic algorithms help the systems to behave like humans—an essential feature that many healthcare systems need to have.

This volume presents seven chapters selected from the rapidly growing application areas of computational intelligence to healthcare systems, including intelligent synthetic characters, man-machine interface, menu generators, analysis of user acceptance, pictures archiving and communication systems, and inverse electromagnetic problem of the heart.

We believe that this volume, along with the first volume of the book, will serve as a useful resource for the health professionals, professors, students, and the computer scientists, who are working on or interested in learning healthcare systems, to overview the current stat-of-the-art of diverse applications of computational intelligence to healthcare practice.

We are grateful to the authors and the reviewers for their vision and great contributions to this book. We are indebted to Springer-Verlag for their excellent help in the preparation of the camera ready copy.

Editors

Contents

1

Introduction to Computational Intelligence in Healthcare

H. Yoshida, S. Vaidya, and L.C. Jain

Abstract. This chapter presents introductory remarks on computational intelligence in healthcare practice, and it provides a brief outline for each of the succeeding chapters in the remainder of this book.

1.1 Computational Intelligence and Healthcare Practice

Computational intelligence provides considerable promise for advancing many aspects of healthcare practice, including clinical disease management such as prevention, diagnosis, treatment, and follow-up, as well as administrative management of patients such as patient information and healthcare delivery to patients.

Computational intelligence is the study of the design of intelligent agents. An intelligent agent is a system that acts intelligently—it does what it thinks appropriate for its circumstances and its goal, it is flexible to changing environments and changing goals, it learns from experience, and it makes appropriate choices given perceptual limitations and finite computation.

However, computational intelligence is more than just the study of the design of intelligent agents, in particular, in application domains. It also includes the study of problems for which there are no effective algorithms, either because it is not possible to formulate them or because they are NP-hard and thus not effective in real life applications. Human being (or biological organisms) can solve such problems every day with various degrees of competence: extracting meaning from perception, understanding language, and solving ill-defined computer vision problems. Thus, the central scientific goal of computational intelligence is to understand the principles that make intelligent behavior possible, whether in natural or in artificial systems.

The central engineering goal of computational intelligence is to specify methods for the design of useful, intelligent artifacts. Indeed, the core methods of computational intelligence—neural computing, fuzzy systems, and evolutionary computing—have recently emerged as promising tools for

H. Yoshida et al.: *Introduction to Computational Intelligence in Healthcare*, Studies in Computational Intelligence (SCI) **65**, 1–4 (2007)
www.springerlink.com

the development, application, and implementation of intelligent systems in healthcare practice. These computational intelligence tools have offered many advantages in automating and creating a physician-like capability in healthcare practice, as demonstrated by the chapters in this book.

1.2 Chapters Included in this Book

The remainder of this book consists of the following seven chapters: Chapter 2 by Hubal et al. reviews several synthetic character technologies pertinent to healthcare applications. Successful development of such technologies requires fusion of various computational intelligence approaches including behavior modeling, natural language interaction, and visualization. This chapter describes pros and cons among different types of synthetic character technologies, surveys the broad range of health-related applications using synthetic characters, and addresses in detail the development and use of synthetic character applications in healthcare practice. Then, the authors focus on four clinically significant applications of a synthetic character in healthcare: assessing skills in obtaining informed consent, assessing skills in dealing with trauma patients, training medical students to interact with pediatric patients, and training law enforcement officers to manage encounters with mentally ill consumers. Assessments of the validity, usability, acceptance, and effectiveness of these applications are also discussed.

Chapter 3 by Gaál et al. presents an automated menu generator of web-based lifestyle counselling systems based on a well-established branch of computational intelligence, genetic algorithms. The menu generator prepares weekly menus that provide users with personalized advice for preventing cardiovascular diseases. The data used in the menu design as derived from personal medical data combined with nutritional guidelines. A genetic algorithm is used for developing a hierarchical organization and a parallel solution for the generation of dietary menus. The authors demonstrate that the menu generator can successfully create dietary menus that satisfy strict numerical constraints on every nutritional level, indicating that such a system can be useful in practice as an online lifestyle counselling system.

Chapter 4 by Zheng et al. presents evaluation methodologies of healthcare IT applications from a user-acceptance perspective. The authors review the theoretical background of intention models that have been widely used for studying factors governing IT acceptance, with particular focus on the technology acceptance model (TAM), which is a prevalent technology adoption theory in the area of information system research. The authors describe the limitations and pitfalls of the TAM, as well as the applicability of the TAM in the professional context of physicians, with a review of available studies that have applied the TAM to technology adoption issues in healthcare practice.

Chapter 5 by Costa et al. presents the current perspectives on picture archiving and communication (PACS) systems pertinent to image-based healthcare practice. PACS-based infrastructures are currently being driven by medical applications that rely on seamless access to medical image databases. The authors review the key factors that have brought PACS technology up to its present status, and they present their web-based PACS as an example of a state-of-the-art system for cardiology services. New, demanding applications such as content based retrieval, computer-aided diagnosis, image-assisted surgery systems, and co-registration among multimodality studies are transforming PACS into a new generation. These future applications of PACS are also discussed in this chapter.

Chapter 6 by Theofilogiannakos et al. presents computational intelligence approaches for inverse electromagnetic problems of the heart. The electrocardiogram (ECG) remains a major tool for the evaluation and management of patients with cardiac disease. Although adequate for management of most patients, there are conditions in which the ECG is suboptimal. For enhancing the diagnostic value of the ECG, the inverse electromagnetic problem needs to be solved; it is defined as the determination of the heart bound from the field that it impresses on the body surface and the geometry of the thorax through which the field spreads. The authors discuss the basic principles of solving the inverse electromagnetic problem of the heart for actual body geometries, as well as the parameters that affect body surface potentials. The authors also present various computational intelligence techniques that are required for obtaining a precise solution of the inverse problem.

Chapter 6.6 by Maggi et al. is regarding a human-machine interface (HMI) for healthcare and rehabilitation. The authors present computational and biomedical approaches for designing an advanced HMI, in particular, direct brain-computer communication. A new miniaturized system is presented for unobtrusive measurement of biological signals using wearable or embedded sensors that are integrated in the advanced HMI. Based on this interface design, a practical brain-computer communication system is designed, which has the promise to provide rehabilitation and healthcare for severely disabled people.

1.3 Conclusion

Advances in computational intelligence have considerable potential to revolutionize healthcare practice. The primary goal of this book is to present some of the most recent research results regarding the applications of computational intelligence to healthcare practice. Readers will gain a wide perspective on this new and rapidly advancing field by reading the present as well as the preceding volume in this book series, Advanced Computational Intelligence Paradigms in Healthcare 1.

Acknowledgements

We are grateful to our numerous colleagues for their contribution during the development phase of this chapter.

References and Further Reading

1. Yoshida, H., Jain, A., Ichalkaranje, A., Jain, L.C., Ichalkaranje, N., Advanced Computational Intelligence Paradigms in Healthcare 1, Springer, 2007
2. Ichalkaranje, N., Ichalkaranje, A. and Jain, L.C., Intelligent Paradigms for Assistive and Preventive Healthcare, Springer, 2006
3. Silvermann, B., Jain, A., Ichalkaranje, A., and Jain, L.C., Intelligent Paradigms in Healthcare Enterprises, Springer, 2005
4. Schmitt, M., Teodorescu, H.N., Jain, A., Jain, A., Jain, S. and Jain, L.C., Computational Intelligence Processing in Medical diagnosis, Springer, 2002
5. Jain, A., Jain, A., Jain, S. and Jain, L.C., Artificial Intelligence Techniques in Breast Cancer Diagnosis and Prognosis, World Scientific, 2000

2

Synthetic Characters in Health-related Applications

R. Hubal, P. Kizakevich, and R. Furberg

Digital Solutions Unit, RTI International, Research Triangle Park, NC, USA

Abstract. This chapter introduces synthetic character technologies, surveys the broad range of health-related applications using synthetic characters, and addresses in detail the development and usage of health-related synthetic character applications.

2.1 Introduction

A number of recent training and assessment healthcare applications have incorporated PC-based synthetic characters as integral components. Realistic-looking synthetic characters rendered on the screen respond appropriately to user manipulations of tools such as scopes and monitors. Characters exhibit such behaviors as gesturing, changing emotional state, linguistic knowledge, and effects on underlying physiology. To achieve this level of realism, synthetic character technology involves behavior modeling, natural language interaction, and visualization.

This chapter describes pros and cons among different synthetic character technologies, surveys the broad range of health-related applications using synthetic characters, and addresses in detail the development and usage of health-related synthetic character applications. The focus is on four specific applications: the use of a synthetic character for assessing skills demonstrated by researchers associated with obtaining informed consent; the use of synthetic characters for assessing skills demonstrated by medical first responders in triage situations and in dealing with trauma patients; the use of synthetic characters for training medical students in strategies for managing clinical interactions with pediatric patients and patients who may have been exposed to bioterrorist agents; and the use of a synthetic character for training law enforcement officers in managing encounters with mentally ill consumers. Assessments of the validity, usability, acceptance, and effectiveness of these applications are also discussed.

R. Hubal et al.: *Synthetic Characters in Health-related Applications*, Studies in Computational Intelligence (SCI) **65**, 5–26 (2007)
www.springerlink.com

2.2 Synthetic Characters

Synthetic characters, also called responsive virtual humans [26] and embodied conversational agents [1], are 3D bodies rendered on a screen with whom a user interacts. Though no set definition constitutes what is a synthetic character application, generally such an application features a language processor, a behavior and planning engine, and a visualization component. The language processor accepts spoken, typed, or selected input from the user and maps this input to an underlying semantic representation [16,18,19,53]. The behavior engine accepts semantic content and other input from the user or system and, using cognitive, social, linguistic, physiological, and other models, determines synthetic character behaviors [2,3,6,14,28,41,45]. Behavior is also affected by non-interactive processes, such as autonomic physiological effects, world and environmental conditions, and synthetic character to synthetic character interactions. These behaviors may include recomputation of subgoal states, changes in emotional state, actions performed in the virtual environment, gestures, body movement, or facial expressions to be rendered, and spoken dialog. The visualization component renders the synthetic character and performs gesture, movement, and speech actions. It may also allow imposition of a selection map on the synthetic character, to support selection of macro anatomy (e.g., the forearm) as well as smaller specific anatomy (e.g., an antecubital vein) and regions (e.g., cardiac auscultation sites). In different applications the user may be immersed in a virtual environment or may interact with a synthetic character simply rendered on a monitor with no appended devices.

There may be a number of reasons for an application designer to employ synthetic characters. When reading facial expression and body language is inherent in training, then interaction makes sense with synthetic characters who use gaze, gesture, intonation, and body posture as well as verbal feedback. Acquiring and practicing interaction skills in a safe and supportive environment allows a student to learn flexible approaches critical for performing well under time constraint, information-poor, and other difficult conditions. Repetitive interactions enable the student to engage in deliberate practice needed to become proficient and confident in skills. Example applications that relate specifically to medicine or health include: medical personnel acquiring and practicing clinical interaction skills [11,31,33,58]; providing research participants with informed consent [25]; dialog with a substance abuse coach or within therapeutic sessions for various phobias [15,22,23,36,50,55]; and interrogation and de-escalation training for law enforcement officers interacting with suspects or mentally unstable individuals [28,42].

2.2.1 Interactivity: Types and Platforms

As stated, in different applications a user may interact with a synthetic character in an immersive virtual environment or with one rendered onto a monitor or screen. Further, the type of interaction may entail verbal input and gesture, menu selection or key entry, mouse or key clicks, or virtual or appended tool use.

An immersive system enables more natural interaction with a synthetic character. There are varying levels of immersive systems [40], from those where the user interacts with a synthetic character projected onto a large screen, to those where the user is also instrumented with sensors that measure position, track gaze, or identify gestures [7, 48], to those where the user is literally immersed in a virtual environment, commonly called a CAVE [9]. Some researchers have set up immersive systems for learning communications skills, arguing that the natural interaction facilitates learning. For instance, one system that was tested with medical students and student physician assistants interacting with a life-sized, projected synthetic patient demonstrated a high level of user engagement and reports of a powerful learning experience [31]. Relatedly, in an application focused on the communications involved in tele-home health care, synthetic characters were used to monitor and appropriately respond to patients' emotions [37]. Meanwhile, in the area of cybertherapy, where virtual environments are used to diagnose and treat patients with various disorders, researchers are beginning to consider how interaction with synthetic characters can assist or improve treatment [47, 50, 55].

Many of the skills that are trained or assessed with synthetic characters, however, do not require immersive systems. For instance, several applications concentrate on communication and strategic skills involved in health promotion, using synthetic characters to guide and support desired health behaviors through long-term relationships [5, 10, 39, 49]. In these applications, synthetic characters who are rendered right on a personal computer monitor or on a mobile platform engage the patient in conversations involving topics such as health nutrition, recognition of heart attack symptoms and management, fitness, and coping with stresses that result from a child's cancer. The synthetic characters in these environments are sufficient as designed to demonstrate empathy, communicate intelligently based on past conversations, and engage the patients in lengthy relationships.

For applications that train or assess procedural skills, generally more realistic environments are necessary but non-immersive environments can be used. That is, though the procedures may involve motor skills that need to be acquired and practiced in a live or highly realistic environment (e.g., with volunteer patients or manikins), learning the procedure itself or assessing a user's procedural activities can often be accomplished in a virtual environment. Likewise, learning when to perform the procedure may be as important as the procedural skill. This is particularly true within team training, where a decision-maker directs others to perform procedures. Virtual environments, ranging from individual computer to distributed multiplayer gaming environments to CAVEs, are quite useful for such training. Some of the applications described in the next section involve the use of specialized tools on synthetic characters. Selection maps are imposed on the synthetic characters, so that procedures that involve use of stethoscopes and otoscopes, application of bandages, checking of pulse or breathing or bleeding, operating a heart monitor or defibrillator, insertion of catheters and needles, and other appropriate medically-related activities (all accomplished by combinations of menu selection, key presses, mouse movement, and mouse clicks), on the correct locations on the synthetic character's body, can be trained or assessed. In addition, procedures that involve checking of scene safety, counting the number of casualties, following trauma protocol, and transporting patients can be implemented using synthetic characters in non-immersive environments.

Before moving to the description of applications, it is instructive to outline why a synthetic character application designer might choose one platform over another. Table 2.1 shows a comparison of application features across various levels of immersivity for medically-related tasks. For skills that truly require realistic activities, such as movement about a scene and coordination with others, instrumented or immersive systems may make sense if live training is not feasible. This may be true for patient safety [38] or terrorism response [54]. However, for many skills, a non-immersive and non-instrumented system that involves synthetic characters is sufficient. Students can acquire and practice numerous procedural, interactive, spatial, and strategic skills by engaging with synthetic characters and virtual environments rendered on a personal computer or mobile device. The typical advantages gained in portability, distribution, cost, throughput, and deliberate practice often outweigh the loss in realism.

2.2.2 Systematic Approach to Training

A course of instruction should result in skilled students, though not necessarily experts. Expertise has its costs: the deliberate practice and diligence required to achieve expert skill is beyond any course length; along with expertise often comes a narrow ability to apply one's skills; and experts can be made to demonstrate overconfidence in their skills [4]. In contrast, a skilled student is able to apply basic and moderately advanced skills not only in the context under which learning took place, but also in analogous situations under changing contexts. Proficiency, then, implies the ability to apply knowledge and skills.

A systematic, cost-effective approach to developing synthetic character training applications, and for assigning learning to environments as outlined in Table 2.1, employs the familiarization, acquisition, practice, validation (FAPV) model [26]. In this model, declarative knowledge is gained during *familiarization*. Declarative knowledge is factual, cognitive, well understood, basic information about skills and contexts. It can be taught in lectures, learned through reading or basic interactive multimedia instruction, or gained informally. It forms the basis for skills acquisition. Procedural knowledge is of processes, beginning as declarative but gradually becoming automated ('proceduralized'). Procedural abilities are gained during *acquisition*, and they are

Table 2.1. Comparison of synthetic character application features across platforms

Platform	Mobile (e.g., personal digital assistant, handheld computer)	PC (interactive desktop 3D)	Instrumented (use of sensor or appended technology) or massively multiplayer	CAVE / WAVE or live environment
Example task	Ecological momentary assessment [51] with a clinical patient	Interact with virtual clinic, trauma, triage patients	VR for surgical skills	Triage scene coordination
Task type	• Individual	• Individual	• Individual • Collective	• Individual • Collective
Task demands	• Interactive/ communi- cative	• Procedural • Interactive/ communi- cative • Strategic	• Procedural • Spatial • Haptic	• Motor • Strategic • Interactive/ communi- cative
Platform affordances	• Portable • Immediate • Low cost • Ubiquitous	• Well under- stood naviga- tional and interaction controls • Sense of engagement • Distribution flexibility • Low cost/ potential high payback	• Realistic interaction • Sense of immersion • May not be prohibi- tively expensive	• Realism • Sense of pres- ence • Can take advantage of existing facilities
Learning levels	• *In situ* assessment	• Familiari- zation • Acquisition • Practice	• Practice • Validation	• Practice • Validation
Student measures	• Dialog length • Actions taken • Patient outcome	• Key, mouse, cursor selections • Virtual navigation about scene • Language input • Timing • Engagement in simulation	• Viewing angle • Actions taken • Tools selected • Timing • Immersion in simulation • Dexterity in hand/eye coordination	• Physical movement about scene • Actions taken • Timing • Presence in simulation

Cost-effectiveness measures: student throughput, ease of distribution, transfer of learning or situated assessment.

automated during *practice*. With practice, this knowledge becomes routine, a skill (often a motor skill) that demands decreasing exertion to accomplish. By definition, practice requires repeated performance in an environment that alters to reflect performance outcomes. It is efficiently done first in virtual worlds (i.e., those that are not instrumented with sensor technology or appended devices such as specialized joysticks), later in more immersive environments (e.g., hands-on or part-task trainers). In understanding when and how to apply knowledge, and in realizing gaps in knowledge that need to be filled for particular tasks, a student demonstrates strategic knowledge. That is, the student strategically applies declarative or procedural knowledge (or both) and experiences consequences. The *validation* of skills that proves they have been successfully acquired must take place in full-up simulators or using live equipment.

Time spent in learning environments decreases as level of proficiency increases. The most time is spent in a traditional environment (e.g., a classroom) becoming familiarized with the tasks. The least time is spent in highly immersive or live environments validating skills. Less immersive environments enable effective acquisition and practice of skills. Also, there may not be a need to validate all skills. Instead, there may be a need to validate some skills, practice to a lesser degree other skills, and be familiarized with still other skills. For instance, if a set of tasks requires comparable skills, then performance on only a small subset of those tasks needs to be validated. For the remaining subset, familiarization and acquisition should prove sufficient, so that the skills will successfully be applied to them on the job. The identification upfront of analogous skills is important for realizing cost-effectiveness from the mix of learning environments.

2.3 Illustrative Applications

In this section some of the issues are discussed surrounding use of synthetic characters for health-related applications. Assessments of the validity, usability, acceptance, and effectiveness of these applications are also discussed.

2.3.1 Assessment of Informed Consent

Informed consent precedes all other healthcare provider / patient communication, and helps establish trust between the parties. It is mandatory not only for provision of care, but also for human subjects research. It is critical, then, for healthcare providers and researchers to be able to respond appropriately to questions regarding informed consent that patients or research participants might pose. One application to assess researchers' ability to provide informed consent, using a synthetic character, was developed for this purpose [25].

The synthetic character played the role of a potential research participant. The character was concerned with the user's responses to a series of questions covering many of the typical elements of informed consent both for research participation and for healthcare delivery: benefits and compensation, confidentiality, a contact for follow-up questions, duration of the procedure, sponsorship of the research, participant selection procedures, and voluntariness. It was the user's job to provide the participant with all relevant information necessary concerning participation, to promote the participant's comprehension of relevant information, and to ensure the participant's voluntariness to consent. The application captured data on how questions were answered by the user, on how the synthetic character's concerns were addressed, and on the consistency and relevance of provided information.

The application's content and criterion validity derive from how it was developed, with continuous input from subject-matter experts that involved logging their tests of the application, revising language grammars to incorporate their dialog and ensure the synthetic character responded appropriately, and retesting after these changes. The application also follows the type of assessment actually used at the authors' institution for informed consent skills.

The application's usability and acceptability was tested in the field in a limited setting [20], as part of a study being conducted on the health effects of those living and working around the World Trade Center during 9/11. Five trained interviewers for that study practiced responding to informed consent questions using the application. The interviewers interacted with the synthetic character for three to six conversations each. After completing the conversations, they filled out a short instrument used in parallel studies (e.g., [43]) on their familiarity with computers and their impressions of the application. The average ratings fell between "moderately" and "very" for questions that were asked about the application regarding how realistic was the character's behavior, how effective the application could be in preparing someone to provide informed consent, how easy was the application to use, and how enjoyable was the application. In addition to the survey filled out by the interviewers, an observer rated how they interacted with the synthetic character, their level of engagement, emotional tone, body language, comprehension, and verbalizations. The interviewers were moderately to highly engaged, relaxed and even amused by the synthetic character, and moderately to highly talkative with the character. Other measures of interaction were less informative, such as the interviewers' low use of body language, negotiation, and information seeking, but this was to be expected given the relatively few body movements and facial gestures made by the character, and the question-answering rather than information-gathering nature of the conversation.

The application's effectiveness was tested in a study conducted with undergraduate students with no specific knowledge of informed consent procedures [25]. In that study, participants either were or were not given time practicing responses to informed consent questions asked by the synthetic character; those who were not were given additional time to learn the same material. Later, all participants were asked to play the role of a researcher obtaining informed consent from a research participant. Outside observers rated the study participants' responses, and found the responses given by those participants who had practiced with the synthetic character application to be superior to the responses given by those participants who had not had the opportunity to practice with the application.

This application represents one approach to training skills that almost exclusively involve communication with another individual. As with any simulated environment, synthetic character simulations can improve interaction skills training by providing students with more practice time and consistent interaction experiences, in a safe, reliable, modifiable environment. Further, this application and others (e.g., [43]) demonstrate that assessment of critical interaction skills, such as are implicated in informed consent, can and perhaps should be required of healthcare providers and others. A synthetic character simulation provides the gamut of interactivity, context, and measurability that valid skills assessment demands.

2.3.2 Trauma Patient Assessment

Emergency medical training benefits from a realistic environment where a novice first responder can fully appreciate the complexities of immediate, pre-hospital trauma care. Forms of training such as practice with instrumented manikins, moulaged-actor simulations, and mock disaster group exercises all contribute to learning, but they also require logistical setup and high student-to-instructor ratios and lack variability in injuries, scenarios, and the dynamic physiological consequences of trauma and treatment.

A series of synthetic character applications have been developed for trauma patient assessment and care [32, 34]. The applications present scenarios comprising a setting (e.g., city street, military bunker), one or more patients with some trauma condition, and a set of caregiver resources. The caregiver can navigate and survey the scene, interact and converse with the synthetic patient, use medical devices, administer medications, monitor data, and perform interventions. To interact with the patient (e.g., taking a pulse), the user right-clicks on the body region of interest (e.g., the wrist). A menu then appears near the selected region, and the appropriate interaction ('Assess pulse') may be selected. Toolbar buttons present options used to treat the synthetic patient that are relevant to the occupation being trained, such as tools (e.g., scissors, knives) to remove layers of clothing, a medical bag with depleting resources, monitors, intravenous catheters and medications, immobilization devices, and transport. In Fig. 2.1, the caregiver has removed the patient's shirt, and applied a blood pressure cuff and electrocardiogram. To manage the casualty's sucking chest wound, an Asherman Chest Seal has been applied, ventilations are being derived manually via a bag valve mask, and vascular access has been established for administration of a crystalloid solution. Spinal precautions have been taken, as indicated by the placement of a cervical collar, and the use of a long spine board.

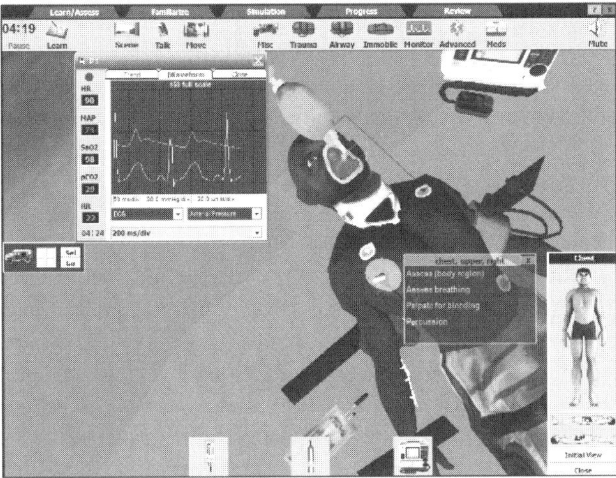

Fig. 2.1. Simulated trauma patient

The applications provide practice of medical care across multiple occupational domains and workplace environments. Since occupations have different assessment and treatment protocols, the user's occupation (e.g., EMT Basic) is specified before simulation to align the learning protocols to the user. Likewise, as described, since work environments have varying resources (i.e., tools, devices, and medications), a resource set is also pre-specified. Thus, in the field, assessment resources may be limited to what can be perceived about the casualty scene with eyes, ears, and senses of touch and smell, and perhaps simple diagnostic tools such as a stethoscope or flashlight. Treatment may be limited to establishing an airway, bandaging bleeding wounds, splinting broken bones, giving intravenous fluids, and transporting the patient. Further, real-life trauma often produces internal injury and physiological complications. The likelihood that such complications will occur during a scenario can be manipulated by an instructor or by the system. In each scenario, the synthetic patient improves, stabilizes, worsens, or dies depending on the care provided. All user interactions are recorded for after-action reviews, as are the pertinent physiological data.

The applications' criterion validity stems from the design to mimic field situations within the constraints imposed by the interface. An ideal trainer would allow the caregiver to acquire and practice all of the skills associated with a level of training. Because an affordable system would not be able to mimic all of the physical attributes of and interactions with a trauma patient, tasks that are primarily cognitive or procedural were separated from those that require motor skills. (The latter demand part-task or hands-on trainers such as manikins that don't require use of screen-rendered synthetic characters.) Learning partitioned this way provides three benefits. First, medical decision making, emergency protocols, physiology, and relative spatial relations between anatomy and medical devices can be learned within a relatively inexpensive simulation. Second, a wide variety of scenarios can be presented to improve decision-making skills. Third, given the learning that takes place in the simulated environment, the caregiver's time needed to validate skills with an expensive, fully-instrumented manikin or with a moulaged standardized patient is reduced.

The applications' content and face validity derives from their continued use of the latest graphics subsystems and modeling capabilities and by use of a real-time physiological model [32,34]. Animations that represent medical signs and procedures, such as vomiting, convulsions, and nausea, are based on motion data captured from instrumented actors playing out the various movements. Facial expressions depict emotion, level of consciousness, reaction to agents, pain, and blink rates. Phonemes are dynamically synchronized with synthetic patient utterances. Casualty visualization for such injuries as a compound fracture or abdominal penetrating wound is shown using dynamic skin texturing. The relations among primary injury, associated injuries, and pathologies are managed, as they are essential for presentation of signs and symptoms and for imposing plausible physiological consequences. Clothing layers (e.g., for a Soldier, underwear, a primary clothing layer, a bulletproof vest, a helmet, socks, boots, and associated gear and hardware), each with its own set of injury textures with appropriate damage or staining depending upon the injury and scene contexts, are loaded and unloaded dependent on application needs, and some patient assessment interactions work differently across layers of clothing. Finally, a physiological model integrates real-time cardiovascular, respiratory, and pharmacokinetic models with state-change conditions such as airway obstruction, pneumothorax, and cardiac tamponade.

One application's usability was tested with students and teachers in emergency medical services programs and practicing emergency medical technicians and paramedics [57]. These users evaluated the accuracy and realism of the applications' organization, tools, visual representations of injuries, caregiver interactions, and physiological responses. Elements of the applications improved through these tests include a more intuitive interface, a greater variety of trauma cases, incorporating additional synthetic characters, and a less distracting tutoring system.

The applications have not yet been shown to be effective. A test with soldiers at an Army medical center yielded inconclusive results. The intent of the study was to validate the use of technology as a training strategy. Specifically, investigators augmented a traditional emergency medical technician curriculum with a trauma care simulator to acquire skills through use of the application and practice in free-play mode. There was some attrition and most soldiers' current assignments were non-medical, however, the study measured differences in passing rates between soldiers who did or did not practice with the application on a written, as opposed to practical, certification exam, so that the transfer of training could be expected to be low in this study design. In addition, overall test scores were used in the comparison between groups of students. Experiment designers failed to account for the weakness of this comparison methodology, as trauma assessment and management questions (the focus of the synthetic character simulations) only comprise approximately 15% of the typical National Registry of Emergency Medical Technicians written examination.

2.3.3 Virtual Standardized Patients

Clinical interaction skills training and assessment can be difficult. Two common approaches are role-playing among students and programs that employ standardized patients to evaluate clinical examination skills of student practitioners. These approaches have limitations, though. For instance, role-playing offers few opportunities to repeatedly practice learned techniques, and provides little or no individualized tutoring. Meanwhile, standardized patient programs are considered effective enough for assessing interaction skills that they are being required for licensure by the National Board of Medical Examiners in the U.S. However, the use of standardized patients is limited by the number of medical students to be assessed and by characteristics (e.g., age, gender, ethnicity, number) of actors available to play the roles of standardized patients [30]. Further, role-plays and standardized patients are useful only for clinical interactions with adult patients and not with pediatric patients. For logistic, reliability, and ethical reasons children are not terrific candidates for role-playing scenarios [27]. For instance, a young child actor would not be expected to behave in a consistent manner across multiple repeated clinical exams for multiple students. Consequently, performance assessment is restricted, and practice opportunities are limited. A final limitation to role-play scenarios involves uncommon but serious clinical presentations, such as possible exposure to bioterrorist agents.

Synthetic characters offer a possible solution, or augmentation to existing assessments, since synthetic character applications may increase the realism of virtual role-plays through natural dialog between the user and emotive, active synthetic characters. Several virtual standardized patient (VSP) systems have been developed as an adjunct to live actors for teaching patient interviewing and history-taking skills, for evaluating students' performance in diagnosing and treating emerging infections and potential incidences of bioterrorism (see Fig. 2.2), and for assessing students' ability to communicate and interact with patients and follow protocol in performing a clinical examination [30, 33].

In a typical scenario, the synthetic patient presents with a chief complaint. The clinician may engage the patient in dialog, consider vital signs, and observe animated or audible cues such as coughing, whining, wheezing, and sneezing. The clinician may make inquiries regarding medical history and physical condition, order diagnostic and laboratory tests, enter differential diagnoses, and plan treatment and patient management. The clinician may need to obtain the trust of the patient before the patient provides accurate or complete responses. Scenarios may take advantage of public health alerts

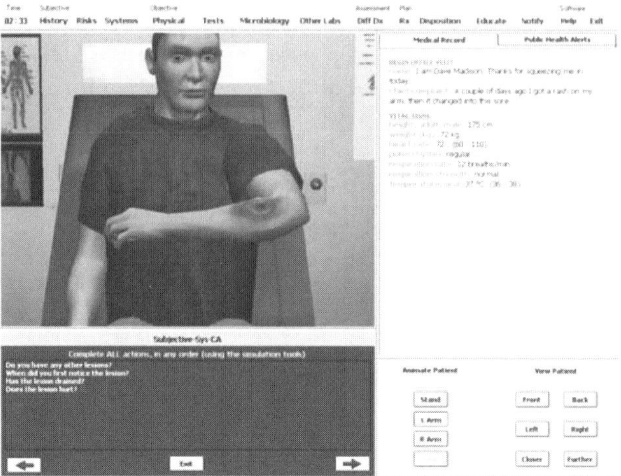

Fig. 2.2. Virtual standardized patient

that should be referenced by the clinician, and may involve presentations of patients having related but more common or not as serious diagnoses (e.g., a spider bite vs. cutaneous anthrax).

These applications' content and criterion validity derive from the reliance on subject-matter expert input for almost all aspects of the simulation. For instance, the scenarios are defined with expert guidance, including the roles, setting, tools available, initial conditions, and scripted events. Similarly, expert input (or findings from a literature review) helps define nearly all of the models that underlie the synthetic character behaviors, as the characters respond to the clinicians' questions and requests and their use of tools. Also, experts provide input for the databases that are needed, such as the day-to-day effects of exposure to different bioterrorist agents and the kinds of activities that a pediatric or adolescent patient would describe during a clinical exam or social history. Finally, experts (including outside experts who are not involved in development) test the applications iteratively to help refine the language, gestural, and other behavioral models that drive synthetic character behavior.

The applications' usability and acceptability was tested in several different venues [11, 27, 33]. The purpose was always to solicit qualitative and quantitative feedback from small groups of potential users in their professional roles as primary care providers. Issues considered in the evaluations were the simulated clinical practice and clinician-patient interactions (including the ability to perform clinical interactions effectively via dialog, menus, buttons, and data summaries), graphics presentation and performance, and freely expressed preference from the program's anticipated users (primary care practitioners). The testing included observation of clinicians engaged in simulation free-play and interviews on the clinicians' reactions to the simulation software. Standard usability testing methods employed included scripted scenarios, data logs, post-test questionnaires, the think-aloud protocol, and test monitor observations; the methods and data collection instruments were adapted from previous usability analyses of synthetic character software [57]. Comments from clinicians centered on visual features of the application (clothes that the synthetic characters wore, realism of the visual presentation of a lesion or wound, usefulness of pictures to show x-rays, gram stains, and the inner ear), manipulation available (e.g., ability to click and drag), and feedback (was the correct diagnosis made, were prescribed medications appropriate, what would have happened to this synthetic patient some number of days later), all implying some level of engagement with the application. Some clinicians even became frustrated when their questions were either not answered or misunderstood, another measure of engagement with the application (but also identification of dialog elements that needed to be, and were, addressed). There was some desire for increased interaction, such as having a nurse available; the consensus was that the clinicians wanted to be able to do more of the actions they might normally take in an actual practice setting. Overall, the applications were rated moderately high to very high, with realistic response time and somewhat realistic objections, concerns, and questions posed by the synthetic characters.

One application's effectiveness has been partly studied. Clinicians on the whole ranked the applications as favorable for use as a training tool. Two scenarios have been developed to a prototype stage for the pediatric VSP application. In one scenario, the clinician is tasked with conducting an ear exam on a very young girl. The girl may be helpful if she is healthy but whiny if she has an ear infection. In another scenario, the clinician must obtain a high-risk behavior history from a teenage girl. Even at this prototype stage, many medical students commented that they learned valuable lessons from interacting with synthetic characters in these scenarios (e.g., not to just move to look right away at a child's ear but to try to establish rapport with the child first, or that a parent needs to be out of the room to get an accurate social history from an adolescent). Importantly, pediatric educators felt these scenarios address pediatric competencies at which only half of their students are competent at graduation. A wide variety of potential audiences was identified who might benefit from using these applications, including emergency physicians, emergency medical technicians, pediatricians, dermatologists, occupational medicine personnel, nurse practitioners, nurses, triage nurses, medical assistants, physician assistants, medical students, and even social workers.

2.3.4 Managing Encounters with the Mentally Ill

Law enforcement officers regularly encounter persons with mental illness but do not always have proper training for these encounters [21, 52]. However, proper training is required, as these encounters can differ markedly from 'normal' encounters, and can lead to adverse events such as incarceration or unnecessary violence [46, 56]. A synthetic character application was developed to support self-paced practice of skills acquired in the classroom [12], where law enforcement officers are trained in the recognition of various types of mental illness and alternative methods (as opposed to their typical use of forceful verbal behavior) of managing encounters with persons exhibiting signs of mental illness or disorders.

With the exception of computer-based training for weapons training and police procedures, lecture tends to dominate law enforcement officer training. However, learning of the interaction skills required to manage an encounter with a person with mental illness, sometimes called 'verbal judo', requires learning by doing, that is, acquiring and practicing skills in an environment that can be expected to lead to transfer of skills to the live environment. This environment requires a virtual role-play with a partner who can exhibit the range of characteristics (age, gender, ethnicity, mental status) that the officer can be expected to encounter on the street. As described, a non-immersive synthetic character application represented a cost-effective approach to training on these interaction skills.

In the synthetic character application, the law enforcement officer encounters a synthetic character, who may or may not be schizophrenic, sitting on or pacing about a bench on a street. The officer's verbal input is analyzed at lexical, syntactic, and semantic levels: specific words or phrases may trigger different behaviors in the synthetic character, forms of phrasing (such as commands vs. requests vs. informational statements) may influence the character's behavior, and the meaning of what was verbalized causes an appropriate response from the synthetic character. The officer's job is to look and listen for indications of particular forms of mental illness so as to adapt responses appropriately, to establish rapport with the synthetic character.

The application's content and criterion validity was established by incorporating, with assistance from an advisory group of experts, a crisis stages model of schizophrenic behavior into the models of behavior of the synthetic character [29]. In addition, a series of tables define how the synthetic character is to behave under varying conditions, including the current emotional state, the current verbal input and topic of conversation, past input, and the current position of the character in the scene [28]. Also, the language models were iteratively tested so that the synthetic character would be able to distinguish among commands, queries, and statements and would respond appropriately based on verbal input and current conditions.

The application's usability and acceptance was evaluated with law enforcement officers taking a course on managing encounters with the mentally ill [12] using both qualitative and quantitative methods. These measures included written pre- and post-test instruments measuring the officer's knowledge of mental illness, course and instructor evaluation, in-class observation, post-test reactions, and discussion groups. The objective of the evaluation was to assess reactions such as perceived utility, ease of use, comfort, and learning enhancement associated with the application. Law enforcement officers who took part represented a mix of occupations, including campus police, sheriff's and municipal law enforcement, traffic enforcement, and mental/medical facility law enforcement. A majority of officers found the simulation easy or somewhat easy to use, rated the simulation as somewhat to very realistic, and felt fairly or extremely comfortable in using the simulation. Officers reflected that the character, true to life, would sometimes not cooperate regardless of what the officer requested. They also commented that the character was dressed relatively nicely, spoke as if he were well-educated, appeared to only recently have been having mental difficulties, and did not represent the extreme fear or dislike of police that officers commonly encounter; these are observations that could not have occurred without engagement with the simulation. Every officer reported that the simulation increased his or her interest in the course and nearly every officer left the course with increased confidence in handling an encounter with a mentally ill person. In an after-class focus group, officers were quite vocal and almost unanimously positive about their experience with the simulation, indicating that they enjoyed its inclusion in the course. In addition to the novelty of the simulation, officers valued the opportunity to work through scenarios independently.

The application's effectiveness within the context of the course was also evaluated. Post-test scores regarding characteristics of persons with mental illness, behaviors, and treatment were significantly higher than pre-test scores, possibly attributable in part to the simulation. Each officer in the course was exposed to the simulation for two periods, first with a partner and later individually. In observing the interactions, many of the officers appeared to carry out constructive and progressive dialogues with the synthetic character, though the observations are somewhat limited in that the observer could only hear one side of the dialogue, and had to intuit what the character was saying or doing. Officers appeared to be very persistent in their interaction; most were calm, courteous and helpful although some officers experimented by using commanding language. The majority of officers indicated that the simulation enhanced their learning in the course to some extent, suggesting its value as a supplement to lecture and citing its utility within a paired learning experience with a partner, as a demonstration tool, and for self-paced interaction.

2.4 Discussion

This section compares the approach taken in creating the synthetic character applications just described with alternative, non-immersive approaches. It also presents possible future directions for synthetic character applications.

2.4.1 Alternative Technologies

Across health-related applications there is no one format for presenting synthetic characters. Some of the different formats include less realistic modeled characters or even character representations, sometimes shown only from the shoulders up (e.g., [5, 6, 23, 39, 45]), realistic modeled characters rendered with a gaming engine, with full body, animations, and morphing capability (e.g., [12, 33, 34, 47]), and video-based or highly realistic characters in limited settings (e.g., [22, 42, 55]). Relatedly, some applications use pre-recorded speech for the synthetic character (e.g., [11, 22, 25, 42, 55]) while others use less realistic but more flexible computer-generated speech or visual display of output text. Less realism generally implies greater modifiability, that is, ease of adapting the characters to have different characteristics or to new settings, and also generally implies ease of use across platforms, such as mobile platforms. More realism generally targets greater buy-in by users into the simulation. However, characters need not be photo-realistic to be engaging and caricatures to be flexible. The level of visual realism combines with convincing linguistic and emotional models to engage users.

Interaction with the synthetic character also varies across applications. For instance, in some applications a natural language processor allows the user to speak naturally with the character (e.g., [22, 25, 30]), in other applications the user is allowed to speak naturally given well-defined conversational topics or enter dialog by typing in free form (e.g., [27, 42]), but in most applications the user converses with the synthetic character using pull-down menus or with constrained text input. Aside from dialog, some applications allow the user to interact with the synthetic character by applying tools to selected regions of the character's body (e.g., [11, 34]). As suggested by the discussion of interactivity, the form of interaction is driven by application demands. Where strictly interaction or communication skills are to be trained or assessed (e.g., [22, 28, 30, 31, 33]), or for applications where the synthetic character is intended to establish a relationship with the user (e.g., [5, 10, 39]), then a form as near to natural language would be desirable for transfer to a live environment. Where procedural skills are to be trained or assessed (e.g., [32, 34]), interaction involving manipulation of the environment would be most desirable. In non-training applications, such as for entertainment, decision support, or research, interaction would be expected to be natural but context-limited.

A number of different behavioral models underlie synthetic characters in different applications. Social models (e.g., [43, 45, 50]) control how the character interprets user actions and how the character responds, relative to the relationship established between the user and character. Physiological models (e.g., [32, 34]) cause the character to change behavior based on tools that the user might employ or based on rules that govern how physiology affects actions. Gestural models (e.g., [2, 6, 28]) and emotional models (e.g., [3, 18]) direct the character's movements, gestures, and expressions. Cognitive models (e.g., [5, 14]) maintain state of knowledge of the world and elements of the interaction. Not every application requires this range of models; the choice of models depends on how complex and how varied behavior the synthetic character needs to exhibit. The architectures used to model behaviors vary across applications, too, from state diagrams to rule bases to goal-directed representations. The application designer's choice of architecture is dependent largely on the behaviors to be simulated [13].

2.4.2 Future Technologies

Use of synthetic characters for health-related applications can be expected to grow, not only into different health-related areas but also by integrating new technologies. The state of the art for future applications will certainly improve over current applications. A few likely advances are presented here.

It is critical in applications for the synthetic character to be able to detect and respond appropriately to poor or inappropriate input from the user, and conversely not to react inappropriately to good input. That is, failing to catch bad responses or not reacting well to good responses breaks the user's flow. In some applications this requirement is met by use of a 'Wizard-of-Oz' approach, whereby a human observer immediately and surreptitiously categorizes the natural language input of a user (e.g., [43]). In most other applications this requirement is met using pull-down menus or similar context-controlling devices. Users will often (but not always intentionally) speak utterances that are outside the range of what the language models expect in the context of the dialog. Further, users will sometimes respond or reply with very complex compound sentences, multiple sentences, and even paragraph-long utterances. If language models are designed well, though, with extensive subject-matter expert input and user testing (perhaps using typed input in the developmental stage in deriving language models [20]), recognition rates for natural language can reach acceptable levels. As an active research area in the computational sciences, natural language processing capabilities can be expected to improve and lead to more realistic synthetic character responses to user input.

More robust and effective behavior models and more efficient means of creating the models, including a modular architecture and interface standards [17], may lead to better behaved characters. Similarly, visual characteristics of modeled characters and activities in the environment (such as models of vehicles, weather, and crowds) continue to improve, largely as a result of gaming technology improvements. Together these advances point towards more realistic applications.

Integrating virtual presentation of casualties with physiologically-responsive patient simulation and haptic casualty interaction using manikins, part-task medical trainers, or other instrumented devices may overcome challenges to sustaining combined cognitive and psychomotor medical skills. For procedural and strategic skills, simulated casualties can be presented using interactive 3D technologies and displayed via a desktop monitor, 2D projection screen, 3D projection with shutter glasses, augmented reality via head-mounted display, or immersive virtual reality via head-mounted display. For haptic interaction, non-instrumented part-task medical trainers (e.g., airway trainers, intravenous arms) have proved their training value for specific haptic tasks, while whole body manikins, with integrated part-task subassemblies, offer greater realism and various levels of instrumentation. Physiologically-responsive casualty simulation for advanced skills training (e.g., triage) can range from bags of 'blood' supporting intravenous training to real-time cardiopulmonary modeling with virtual waveform display, pharmacokinetic modeling, or tongue swelling or vomitus provided in an advanced airway manikin.

Current applications make no use of the user's vocal affect, facial expressions, eye movement, body gesture, or reasonably non-invasive physiological input [8, 35, 44] in interpreting the user's emotional state and intentions. However, another very active area of research, called 'augmented cognition', seeks to enhance the user's experience with a system by sensing the user's current state and adapting the interface to improve decision-making and overcome limitations and biases [48]. Augmented systems for training, using synthetic characters, can be expected to improve the learning experience by better adapting character behaviors to the user. Along similar lines, improved dynamic measures of a user's current knowledge and abilities can be expected to help identify differences against a standard of strategies and procedures exhibited by an expert. That is, better models of the user might lead to more effective or efficient adaptive steps taken by the application and hence to more effective or efficient learning.

2.5 Summary

Procedural and interaction skills training and assessment in health-related contexts often rely on passive forms through hearing or seeing but not often active forms by doing. When skills are trained or assessed using role-playing, students are limited in the practice time and the variety of scenarios that they encounter. A scenario such as managing a pediatric patient in the clinic, or a trauma patient on-site, or a mentally ill patient on the street, is difficult to create in a role-play.

Synthetic character technology links theory of human behavior with virtual environments, knowledge representation, and natural language processing. The form of user interaction can vary from desktop interactive 3D to immersive systems, but in general, in synthetic character applications focused on training and assessment, users employ visual cues as well as react to verbal responses to successfully complete each scenario. For the types of skills these applications target, they have demonstrated usability, accessibility, and cost-effectiveness, and also offer benefits including ease of adaptability, availability, variability, and consistency. Improvements and enhancements to the technology promise additional applications in new health-related areas.

References

1. André, E., and Rehm, M.: Guest editorial. Künstliche Intelligenz (KI Journal), Special Issue on Embodied Conversational Agents 17 (2003) 4
2. André, E., Rist, T., and Müller, J.: Employing AI methods to control the behavior of animated interface agents. International Journal of Applied Artificial Intelligence 13 (1999) 415–448
3. Bates, J.: The role of emotion in believable agents. Communications of the ACM 37 (1994) 122–125
4. Bédard, J., and Chi, M.T.H.: Expertise. Current Directions in Psychological Science 1 (1992) 135–139
5. Bickmore, T., Gruber A, and Picard R.: Establishing the computer-patient working alliance in automated health behavior change interventions. Patient Education and Counseling 59 (2005) 21–30
6. Cassell, J., and Vilhjálmsson, H.H.: Fully embodied conversational avatars: making communicative behaviors autonomous. Autonomous Agents and Multi-Agent Systems 2 (1999) 45–64
7. Cassell, J., Bickmore, T., Billinghurst, M., Campbell, L., Chang, K., Vilhjálmsson, H., and Yan, H.: An architecture for embodied conversational characters. Workshop on Embodied Conversational Characters 1 (1998) 21–30
8. Conati, C.: Probabilistic assessment of user's emotions in educational games. Applied Artificial Intelligence 16 (2002) 555–575
9. Cruz-Neira, C., Sandin, D.J., DeFanti, T.A., Kenyon, R.V., and Hart, J.C.: The CAVE: audiovisual experience automatic virtual environment. Communications of the ACM 35 (1992) 67–72
10. de Rosis, F., Novielli, N., Carofiglio, V., Cavalluzzi, A., and De Carolis, B.: User modeling and adaptation in health promotion dialogs with an animated character. Journal of Biomedical Informatics 39 (2006) 514–531
11. Deterding, R., Milliron, C., and Hubal, R.: The virtual pediatric standardized patient application: formative evaluation findings. Studies in Health Technology and Informatics 111 (2005) 105–107
12. Frank, G., Guinn, C., Hubal, R., Pope, P., Stanford, M., and Lamm-Weisel, D.: JUST-TALK: an application of responsive virtual human technology. Interservice/Industry Training, Simulation and Education Conference 24 (2002) 773–779
13. Fu, D., Houlette, R., and Ludwig, J.: Intelligent behaviors for simulated entities. Interservice/Industry Training, Simulation and Education Conference 27 (2005) 1654–1660
14. Funge, J., Tu, X., and Terzopoulos, D.: Cognitive modeling: knowledge, reasoning and planning for intelligent characters. Computer Graphics and Interactive Techniques 26 (1999) 29–38
15. Gaggioli, A., Mantovani, F., Castelnuovo, G., Wiederhold, B., and Riva, G.: Avatars in clinical psychology: a framework for the clinical use of virtual humans. CyberPsychology & Behavior 6 (2003) 117–125
16. Godéreaux, C., El Guedj, P.O., Revolta, F., and Nugues, P.: A conversational agent for navigating in virtual worlds. Humankybernetik 37 (1996) 39–51
17. Gratch, J., Rickel, J., André, E., Badler, N., Cassell, J., and Petajan, E.: Creating interactive virtual humans: some assembly required. IEEE Intelligent Systems 17 (2002) 54–63

18. Guinn, C., and Hubal, R.: Extracting emotional information from the text of spoken dialog. User Modeling 9 (2003) Workshop on Assessing and Adapting to User Attitudes and Affect: Why, When and How?
19. Guinn, C.I., and Montoya, R.J.: Natural language processing in virtual reality. Modern Simulation & Training 6 (1998) 44–55
20. Guinn, C., Hubal, R., Frank, G., Schwetzke, H., Zimmer, J., Backus, S., Deterding, R., Link, M., Armsby, P., Caspar, R., Flicker, L., Visscher, W., Meehan, A., and Zelon, H.: Usability and acceptability studies of conversational virtual human technology. SIGdial Workshop on Discourse and Dialogue 5 (2004) 1–8
21. Hails, J., and Borum, R.: Police training and specialized approaches to respond to people with mental illnesses. Crime & Delinquency 49 (2003) 52–61
22. Harless, W.G., Zier, M.A., Harless, M.G., and Duncan, R.C.: Virtual conversations: an interface to knowledge. IEEE Computer Graphics and Applications 23 (2003) 46–52
23. Hayes-Roth, B., Amano, K., Saker, R., and Sephton, T.: Training brief intervention with a virtual coach and virtual patients. Annual Review of CyberTherapy and Telemedicine 2 (2004) 85–96
24. Hemman, E.A.: Improving combat medic learning using a personal computer-based virtual training simulator. Military Medicine 170 (2005) 723–727
25. Hubal, R.C., and Day, R.S.: Informed consent procedures: an experimental test using a virtual character in a dialog systems training application. Journal of Biomedical Informatics 39 (2006) 532–540
26. Hubal, R., and Frank, G.: Interactive training applications using responsive virtual human technology. Interservice/Industry Training, Simulation and Education Conference 23 (2001) 1076–1086
27. Hubal, R.C., Deterding, R.R., Frank, G.A., Schwetzke, H.F., and Kizakevich, P.N.: Lessons learned in modeling pediatric patients. Studies in Health Technology and Informatics 94 (2003) 127–130
28. Hubal, R.C., Frank, G.A., and Guinn, C.I.: Lessons learned in modeling schizophrenic and depressed responsive virtual humans for training. Intelligent User Interfaces 7 (2003) 85–92
29. Hubal, R., Frank, G., Guinn, C., and Dupont, R.: Integrating a crisis stages model into a simulation for training law enforcement officers to manage encounters with the mentally ill. AAAI Spring Symposium Series Workshop on Architectures for Modeling Emotion: Cross-Disciplinary Foundations (2004) 68–69
30. Hubal, R.C., Kizakevich, P.N., Guinn, C.I., Merino, K.D., and West, S.L.: The virtual standardized patient: simulated patient-practitioner dialogue for patient interview training. Studies in Health Technology and Informatics 70 (2000) 133–138
31. Johnsen, K., Dickerson, R., Raij, A., Harrison, C., Lok, B., Stevens, A., and Lind, D.S.: Evolving an immersive medical communication skills trainer. Presence: Teleoperators and Virtual Environments 15 (2006) 33–46
32. Kizakevich, P.N., Duncan, S., Zimmer, J., Schwetzke, H., Jochem, W., McCartney, M.L., Starko, K., and Smith, N.T.: Chemical agent simulator for emergency preparedness training. Studies in Health Technology and Informatics 98 (2004) 164–170
33. Kizakevich, P.N., Lux, L., Duncan, S., Guinn, C., and McCartney, M.L.: Virtual simulated patients for bioterrorism preparedness training. Studies in Health Technology and Informatics 94 (2003) 165–167

34. Kizakevich, P.N., McCartney, M.L., Nissman, D.B., Starko, K., and Smith, N.T.: Virtual medical trainer: patient assessment and trauma care simulator. Studies in Health Technology and Informatics 50 (1998) 309–315
35. Kizakevich, P.N., Teague, S.M., Nissman, D.B., Jochem, W.J., Niclou, R., and Sharma, M.K.: Comparative measures of systolic ejection during treadmill exercise by impedance cardiography and Doppler echocardiography. Biological Psychology 36 (1993) 51–61
36. Klinger, E., Bouchard, S., Légeron, P., Roy, S., Lauer, F., Chemin, I., and Nugues, P.: Virtual reality therapy versus cognitive behavior therapy for social phobia: a preliminary controlled study. CyberPsychology & Behavior 8 (2005) 76–88
37. Lisetti, C.L., Nasoz, F., Lerouge, C., Ozyer, O., and Alvarez, K.: Developing multimodal intelligent affective interfaces for tele-home health care. International Journal of Human-Computer Studies 59 (2003) 245–255
38. Liu, A., and Bowyer, M.: Patient safety and medical simulation: issues, challenges and opportunities. Medicine Meets Virtual Reality 14 (2006)
39. Marsella, S., Johnson, W.L., and LaBore, C.M.: Interactive pedagogical drama for health interventions. International Conference on Artificial Intelligence in Education 11 (2003) 341–348
40. Milgram, P., and Kishino, F.: A taxonomy of mixed reality visual displays. IEICE Transactions on Information Systems E77-D (1994) 1321–1329
41. Norling, E., and Ritter, F.E.: Towards supporting psychologically plausible variability in agent-based human modelling. Autonomous Agents and Multiagent Systems 3 (2004) 758–765
42. Olsen, D.E., and Sticha, D.: Interactive simulation training: computer simulated standardized patients for medical diagnosis. Studies in Health Technology and Informatics 119 (2006) 413–415
43. Paschall, M.J., Fishbein, D.H., Hubal, R.C., and Eldreth, D.: Psychometric properties of virtual reality vignette performance measures: a novel approach for assessing adolescents' social competency skills. Health Education Research: Theory and Practice 20 (2005) 61–70
44. Picard, R.W., Vyzas, E., and Healey, J.: Toward machine emotional intelligence: analysis of affective physiological state. IEEE Transactions on Pattern Analysis and Machine Intelligence 23 (2001) 1175–1191
45. Prendinger, H., and Ishizuka, M.: Social role awareness in animated agents. Autonomous Agents 5 (2001) 270–277
46. Price, M.: Commentary: the challenge of training police officers. Journal of the American Academy of Psychiatry and the Law 33 (2005) 50–54
47. Rizzo, A.A., Bowerly, T., Shahabi, C., Buckwalter, J.G., Klimchuk, D., and Mitura, R.: Diagnosing attention disorders in a virtual classroom. IEEE Computer 37 (2004) 87–89
48. Schmorrow, D.D. (Ed.): Foundations of augmented cognition. Mahwah, NJ: Lawrence Erlbaum Associates (2005)
49. Silverman, B.G., Holmes, J., Kimmel, S., and Branas, C.: Computer games may be good for your health. Journal of Healthcare Information Management 16 (2002) 80–85
50. Slater, M., Pertaub, D.P., and Steed, A.: Public speaking in virtual reality: facing an audience of avatars. IEEE Computer Graphics and Applications 19 (March/April 1999) 6–9

51. Smyth, J., and Stone, A.: Ecological momentary assessment research in behavioral medicine. Journal of Happiness Studies 4 (2003) 35–52
52. Steadman, H.J., Deane, M.W., Borum, R., and Morrissey, J.P.: Comparing outcomes of major models for police responses to mental health emergencies. Psychiatric Services 51 (2000) 645–649
53. Stokes, J.: Speech interaction and human behavior representations (HBRs). Computer Generated Forces and Behavioral Representation 10 (2001) 467–476
54. Swift, C., Rosen, J.M., Boezer, G., Lanier, J., Henderson, J.V., Liu, A., Merrell, R.C., Nguyen, S., Demas, A., Grigg, E.B., McKnight, M.F., Chang, J., and Koop, C.E.: Homeland security and virtual reality: building a strategic adaptive response system (STARS). Studies in Health Technology and Informatics 111 (2005) 549–555
55. Takács, B. (2005), Special education & rehabilitation: teaching and healing with interactive graphics. IEEE Computer Graphics and Applications 25 (September/October 2005) 40–48
56. Teplin, L.A.: Keeping the peace: police discretion and mentally ill persons. National Institute of Justice Journal 244 (2000) 8–15
57. Weaver, A.L., Kizakevich, P.N., Stoy, W., Magee, J.H., Ott, W., and Wilson, K.: Usability analysis of VR simulation software. Studies in Health Technology and Informatics 85 (2002) 567–569
58. Ziemkiewicz, C., Ulinski, A., Zanbaka, C., Hardin, S., and Hodges, L.F.: Interactive digital patient for triage nurse training. International Conference on Virtual Reality 1 (2005)

3

Application of Artificial Intelligence for Weekly Dietary Menu Planning

Balázs Gaál, István Vassányi, and György Kozmann

University of Pannonia, Department of Information Systems
H-8200 Veszprém, Hungary {bgaal, vassanyi, kozmann}@irt.vein.hu

Abstract. Dietary menu planning is an important part of personalized lifestyle counseling. The chapter describes the results of an automated menu generator (MenuGene) of the web-based lifestyle counseling system Cordelia that provides personalized advice to prevent cardiovascular diseases. The menu generator uses Genetic Algorithms to prepare weekly menus for web users. The objectives are derived from personal medical data collected via forms, combined with general nutritional guidelines. The weekly menu is modeled as a multi-level structure. Results show that the Genetic Algorithm based method succeeds in planning dietary menus that satisfy strict numerical constraints on every nutritional level (meal, daily basis, weekly basis). The rule-based assessment proved capable of manipulating the mean occurrence of the nutritional components thus providing a method for adjusting the variety and harmony of the menu plans. By splitting the problem into well determined subproblems, weekly menu plans that satisfy nutritional constraints and have well assorted components can be generated with the same method that is used for daily and meal plan generation.

3.1 Introduction

The Internet is a common medium for lifestyle counseling systems. Most systems provide only general advice in a particular field, others employ forms to categorize the user, in order to give more specific information. They also often contain interactive tools for menu planning [1].

The aim of the Cordelia project [2] is to promote the prevention of Cardiovascular Diseases (CD), identified as the leading cause of death in Hungary, by providing personalized advice on various aspects of lifestyle, an important part of which is nutrition.

MenuGene, the automated menu planner integrated with Cordelia uses the computational potential of today's computers which offers algorithmic solutions to hard problems. The quality of these computer made solutions may be lower than those of qualified human professionals, but they can be computed on demand and in unlimited quantities. Nutrition counseling is one of this

B. Gaál et al.: *Application of Artificial Intelligence for Weekly Dietary Menu Planning*, Studies in Computational Intelligence (SCI) **65**, 27–48 (2007)
www.springerlink.com

kind of problems. Human professionals possibly surpass computer algorithms in quality, although research comparing performance has been ongoing since the 1960's.

The core idea of our algorithm is the hierarchical organization and parallel solution of the problem. Through the decomposition of the weekly menu planning problem, nutrient constraints can be satisfied on the level of meals, daily plans and weekly plans simultaneously. This feature, which is a novelty, makes the implementation of our method instantly applicable in practice.

3.2 Objectives of Nutrition Counseling

There is no generally accepted method for producing a good menu. Additionally, a menu plan, whether it is weekly, daily or single meal, can only be evaluated when it is fully constructed. So the basic objective of our work is to design a menu planner which also includes some method to evaluate menu plans.

3.2.1 Evaluation of Dietary Plans

The evaluation of a meal plan has at least two aspects. Firstly, we must consider the quantity of nutrients. There are well defined constraints for the intake of nutrient components such as carbohydrate, fat or protein which can be computed for anyone, given their age, gender, body mass, type of work, age and diseases. Optimal and extreme values can be specified for each nutrient component. So as for quantity, the task of planning a meal can be formulated as a constraint satisfaction and optimization problem.

Secondly, the harmony of the meal's components should be considered. Plans satisfying nutritional constraints should also be appetizing. The dishes of a meal should go together. By common sense some dishes or nutrients do not appeal in the way others do. This common sense of taste and cuisine should be incorporated in any nutritional counselor designed for practical use.

There could also be conflicting numerical constraints or harmony rules. A study found that even menus made by professionals may fail to satisfy all of the nutrient constraints [3].

3.2.2 Personalized Objectives

The information collected via web forms in Cordelia explores controllable and uncontrollable risk factors for CD. Controllable risk factors include smoking, high blood pressure, diabetes, high cholesterol level, obesitas, lack of physical activity, stress and oral contraceptives. Uncontrollable factors considered are age, gender and family CD history. Based on the answers, the user is classified, the classification being a combination of factors like overweight, high cholesterol, etc.

MenuGene uses the user's classification and all other useful observations (like the gender) and personal preferences (set by the user) to plan daily and weekly menus. This information is used to design the actual runtime parameters (objectives) of the menu to be generated when MenuGene is run. The nutritional allowances are looked-up from a table similar to Dietary Reference Intakes (DRI) [4, 5].

The fact base of MenuGene was loaded with the data of a commercial nutritional database, developed especially for Hungarian lifestyle and cuisine, that at present contains the recipes of 569 dishes with 1054 ingredients. The database stores the nutritional composition of the ingredients. The recipes specify the quantity of each ingredient in the meal, so the nutrients of a meal can be calculated by summation. At present, the nutrients contained in the database for each ingredient are energy, protein, fat, carbohydrates, fiber, salt, water, and potassium. Additionally, the database contains the categorization of the ingredient as either of the following: cereal, vegetable, fruit, dairy, meat or egg, fat and candy. This classification is used by MenuGene to check whether the overall composition (with respect to the ratio of the categories) conforms to the recommendations of the "food pyramid".

3.3 State of Art in Nutrition Counseling Expert Systems

Research in the field of computer aided nutrition counseling has begun in the 1960s. In 1964 Balintfy developed a linear programming method for optimizing menus [6]. Balintfy's computer code has been developed to plan menus by finding minimal cost combinations of menu items such that the daily dietary, gastronomic and production requirements can be satisfied for a sequence of days. While the menus met nutritional constraints, they did not satisfy the aspects of harmony. From an economical point of view, according to the author, up to 30% percent of food cost saving was possible with the software, however a considerable amount of data processing had to precede the implementation of the system. The code was written in FORTRAN and was running on the IBM 1410 computer.

In 1967 Eckstein used random search to satisfy nutritional constraints for meals matching a simple meal pattern. Each menu was composed of seven components which were meat, starchy food, vegetable, salad, dessert, beverage and bread. Food items were randomly chosen for each component and were evaluated by criteria including calories, cost, color and variety [7]. The program was iterated until the menu generated was satisfactory.

Later, artificial intelligence methods were developed mostly using Case-Based Reasoning (CBR) or Rule-Based Reasoning (RBR) or combining these two with other techniques [8]. A hybrid CBR-RBR system, CAMPER [9] integrates the advantages of the two independent implementations: the case-based menu planner, CAMP [10] and PRISM [11]. CAMP is a case-based reasoner, its case base holds 84 daily menus obtained from recognized nutrition sources, reviewed by experts for adequacy and modified to ensure that each one conforms to the Reference Daily Intakes and aesthetic standards. However, no one menu is good for all individuals, as they may vary in their tastes and nutrition needs. CAMP stores solutions for daily menus and also records their usefulness which is computed according to the menus' nutrient vector, the types of meals and number of snacks and foods included. CAMP operates by retrieving and adapting daily menus from its case base. CAMP's adaptation framework is based on the manual approach of nutritional experts to design menus. Meal level and food level variations are performed before nutrient level adaptations. A detailed description of CAMP is given in [10]. The rule-based menu planner PRISM performs the same task as CAMP but in a different fashion. PRISM relies on menu and meal patterns and its approach to menu creation is generate, test, and repair. A daily menu is generated by successively refining patterns of meals, dishes, foods, and filling general pattern slots. After a menu is generated to fit both user specifications and common sense, it is tested to see if it meets nutritional constraints. PRISM uses a backtracking process to repair solutions, in which new foods, dishes or meals are substituted for those found to be nutritionally lacking. CAMPER is an integration of the techniques employed by PRISM and CAMP. The CBR module was taken intact from CAMP and the RBR module was modeled on PRISM. The database of CAMPER was also more sophisticated than those of its predecessors, containing data for 608 food items, and describing each role a food can fulfill. According to the authors, CBR and RBR complement each other in CAMPER. CBR contributes an initial menu that meets design constraints by building on food combinations that have proven satisfactory in the past, and RBR allows the analysis of alternatives, so that innovation becomes possible. The ability to produce new cases for later use by a CBR module is significant. This enables the system to improve its performance over time. CAMPER, taking advantage of CBR/RBR synergy, provides a capability which neither CAMP nor PRISM provided.

A more recent CBR approach is MIKAS, menu construction using incremental knowledge acquisition system [12]. MIKAS allows the incremental development of its knowledge-base. Whenever the results are unsatisfactory, an expert will modify the system-produced diet manually. MIKAS asks the expert for simple explanations for each of the manual actions he/she takes and incorporates the explanations into its knowledge base [13]. Also, a web based system entitled DIETPAL that models the workflow of dietitians has been built in Malaysia recently for dietary menu generation and management [14]. According to the authors, the main novelty of their system is the use of the complete dietary-management system currently adhered by dietetians in Malaysia. In addition, Dietpal is implemented as a Web-based application, therefore the outreach of the system for use by dietetians and health professionals within the same hospital or at other locations is increased. The system is also capable of storing and organizing patients' dietary records and other health-diet related information, which allows dietetians to effectively evaluate and monitor the patients' dietary changes throughout the period of consultations.

While more than a few expert systems have been developed recently for nutrition counseling, a solution that at least tries to satisfy each and every aspect of an ideal menu planning is still missing. We believe that an effective solution that could support the various cuisines, eating habits, user preferences and other criteria, can only be made by applying general, preferably problem independent algorithms.

3.4 The Basic Algorithm of MenuGene

MenuGene uses Genetic Algorithms for the generation of dietary plans. A genetic algorithm (GA) is an algorithm often used for the solution of "hard" problems and it is based on the principles of evolutionary biology and computer science. Genetic algorithms use techniques such as inheritance, mutation, natural selection and recombination derived from biology. In GAs a population of abstract representations of candidate solutions (also called chromosomes, genomes or individuals) evolves toward better solutions. The evolution starts with a population containing random individuals and happens in generations, in which stochastically selected individuals are modified (via recombination or mutation) to form the population of the next iteration. The attributes (also called alleles) of the chromosomes contain the information where each attribute represents a property. Genetic algorithms are used widely in the medical field [15–18].

GAs showed their strength in satisfying optimization problems, therefore we examined their efficiency in the generation of meal plans. A test software was developed to analyze the adequacy of GAs. In the followings, we describe the basic operators and the fitness function adopted. Experiments with these operators are described in Sect. 7.4.

3.4.1 Genetic Operators

In order to start the genetic search process, we first need an initial population. This may be created randomly or may be loaded from a data base containing solutions of similar cases (Case Based Reasoning, CBR). The population in our tests contained 40 to 200 individuals, which are meals if we plan a single meal, daily menus if we plan a daily menu etc.

In the case of a meal plan, the population contains meals, the attributes of which are dishes. Then, in each iteration step, we execute a sequence of the evolutionary operators (crossover, mutation, selection) on the individuals. We stop the evolution process after a maximum of 1000 cycles (generations)

Table 3.1. The crossover operator is shown in the table in function of the nutritional level

level	● parent 1	○ parent 2	◐ offspring 1	◑ offspring 2	cp
weekly	M Tu W Th F Sa Su	M Tu W Th F Sa Su	M Tu W Th F Sa Su	M Tu W Th F Sa Su	3
daily	BF MS L AS D	BF MS L AS D	BF MS L AS D	BF MS L AS D	3
meal	S G T Dr De	S G T Dr De	S G T Dr De	S G T Dr De	4
abstract	◇ ◇ ◇ ◇	◇ ◇ ◇ ◇	◇ ◇ ◇ ◇	◇ ◇ ◇ ◇	2

Legend: M – Monday, Tu – Tuesday, W – Wednesday, Th – Thursday, F – Friday, Sa – Saturday, Su – Sunday BF – Breakfast, MS – Morning Snack, L – Lunch, AS – Afternoon Snack, D – Dinner S – Soup, G – Garnish, T – topping, Dr – drink, De – Dessert, cp – Crossover Point

or when no significant improvement could be achieved. The best individual of the final population is selected as the solution.

The evolutionary operators are presented via the following example. A regular Hungarian lunch consists of five parts: a soup (1), a main dish consisting of a garnish (2) e.g. mashed potatoes and a topping (3) e.g. a slice of meat, a drink (4), and a dessert (5). So, a solution for a regular Hungarian lunch contains five attributes. The crossover operator acts upon two solutions and it means that starting at a random point, their attributes are swapped. For example, if the starting point is the last attribute, then crossover means the exchange of the desserts. Mutation replaces a randomly selected dish with another one of the right sort (e.g. a soup with another soup).

The single point crossover (recombination) operator is exemplified in Table 3.1. Recombination is done by randomly choosing a crossover point (cp) and creating two offsprings by exchanging the attributes of the solutions from that point on. On the weekly level, the attributes of the solutions represent daily menu plans. In the example in Table 3.1, we apply crossover to weekly level solutions, with a randomly chosen crossover point (cp = 3). The first offspring will contain the daily menu plans for Monday, Tuesday and Wednesday from the first parent and Thursday, Friday, Saturday and Sunday from the second parent. The genetic operators work on the abstract solution and attribute classes and do not operate on problem specific data, thus the same method is used on every level.

In our test system MenuGene, solutions already found for a problem type (for example: low cholesterol breakfast plans) can be reused by loading them to the appropriate initial populations. Therefore, we store some of the best solutions found for each problem. Whenever MenuGene is about to create a new plan that must satisfy certain constraints, it searches its database for solutions that were generated with similar constraints, and loads them as an initial/startup population.

3.4.2 The Fitness Value

In any genetic algorithm, every time a new individual (offspring) is created by mutation or recombination, the fitness function assesses it according to its goodness. In a meal plan, "goodness" means the right physical amount of ingredients and the harmony of the dishes making up the meal. The first aspect can be controlled via numerical constraints, preferably personalized for the user, but the latter requires expert knowledge and will be elaborated in Sect 3.6.

As for numerical constraints, the fitness function has to discard solutions having inadequate amount of nutrients. In our system, the *fitness* of a solution is defined by the sums of functions composed of four quadratic curves (penalty functions) that take their maximum (which is 0) at the specified optimum parameters, and break down abruptly over the upper and lower limit parameters (see Fig. 3.1). For example, if a set of constraints (upper and lower limit, optimal value) is defined for carbohydrate and protein separately, then the fitness is a sum of the two values taken from the two penalty functions for the carbohydrate and protein curves at the respective amounts. The actual fitness value bears no concrete physical meaning; it is used only for comparison and selection. The individuals with the highest (i.e. closest to zero) fitness values are considered the best solutions for the search problem.

The penalty function is thus designed that it should not differentiate small deviations from the optimum but be strict on values that are not in the interval defined by the constraints. The function is not necessarily symmetric to the optimum, because the effects of the deviations from the optimal value can be different in the negative and positive case. This sort of penalizing has been derived from the manual assessment methods of human nutrition experts.

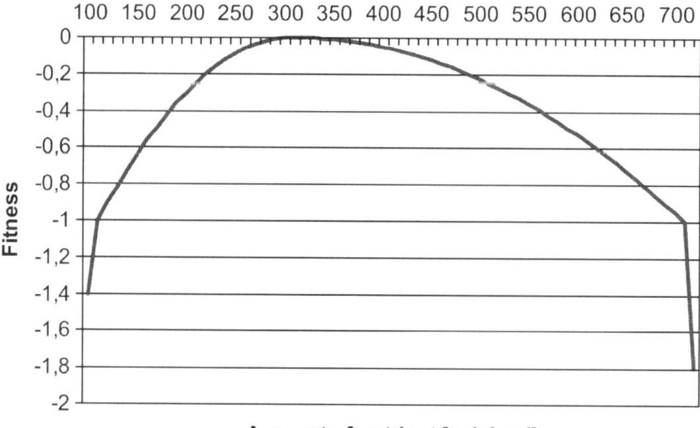

Fig. 3.1. The Fitness function with optimum = 300, lower limit = 100, upper limit = 700

Fitness functions of this style are also often applied in other multi-objective optimization techniques [19].

3.5 Multi-level Organization

There are several common features in the different nutritional levels (meal plan, daily plan, weekly plan) of the problem space. For example, both a meal and a daily plan can be considered a solution of a GA, the attributes of daily plans being meals, and the attributes of meals being dishes (see Fig. 3.2). This feature is the basis for a divide-and-conquer approach we used to design the data structures and the scheduling policy of our test software framework.

3.5.1 Multi-level Data Structures

The problem of generating weekly menus can be divided into sub-problems, which in turn can be solved the same way using GAs. Recently a similar approach, a multi-level GA was presented and tested on a multi-objective optimization problem [20]. In our approach we hope to develop and enhance this idea. In our test system MenuGene, we created a C + + framework called GSLib, which uses the functions provided by the GALib [21] genetic algorithm library for running a standard, parameterized evolution process on the current population. GSLib is abstracted from the menu generation problem and uses abstract classes such as "solution" and "attribute" to represent the information related to the optimization and constraint satisfaction problems. GSLib is used to initialize the algorithm, to control the multi-level divide/conquer style scheduling, and to operate the abstract evolution processes at various levels.

Thanks to the abstract framework, every kind of meal can be represented and every kind of plan can be generated the same way, irrespective of national cuisine, eating habits and nutrition database. For example, in Hungarian hospitals, "hot dinners" i.e. those containing at least a hot soup, are served two times a week. A solution type for this kind of weekly plan can be easily recorded in our database. It would contain five "regular daily plan with cold dinner" attributes for every day except Thursday and Sunday where we would have the attributes for "regular daily plan with hot dinner". Another example would be a school cafeteria, where only breakfasts, morning snacks and lunches are served. Five attributes, only for weekdays, for "daily plan for school refectories" would make up the solution for a weekly plan.

3.5.2 Multi-level Scheduling Policy

Weekly menu plans can be generated in sequential form. Meal-level GAs could create meals from dishes, then these meals could be used as a fact base to generate daily plans. The same applies for weekly menu plan generation. However,

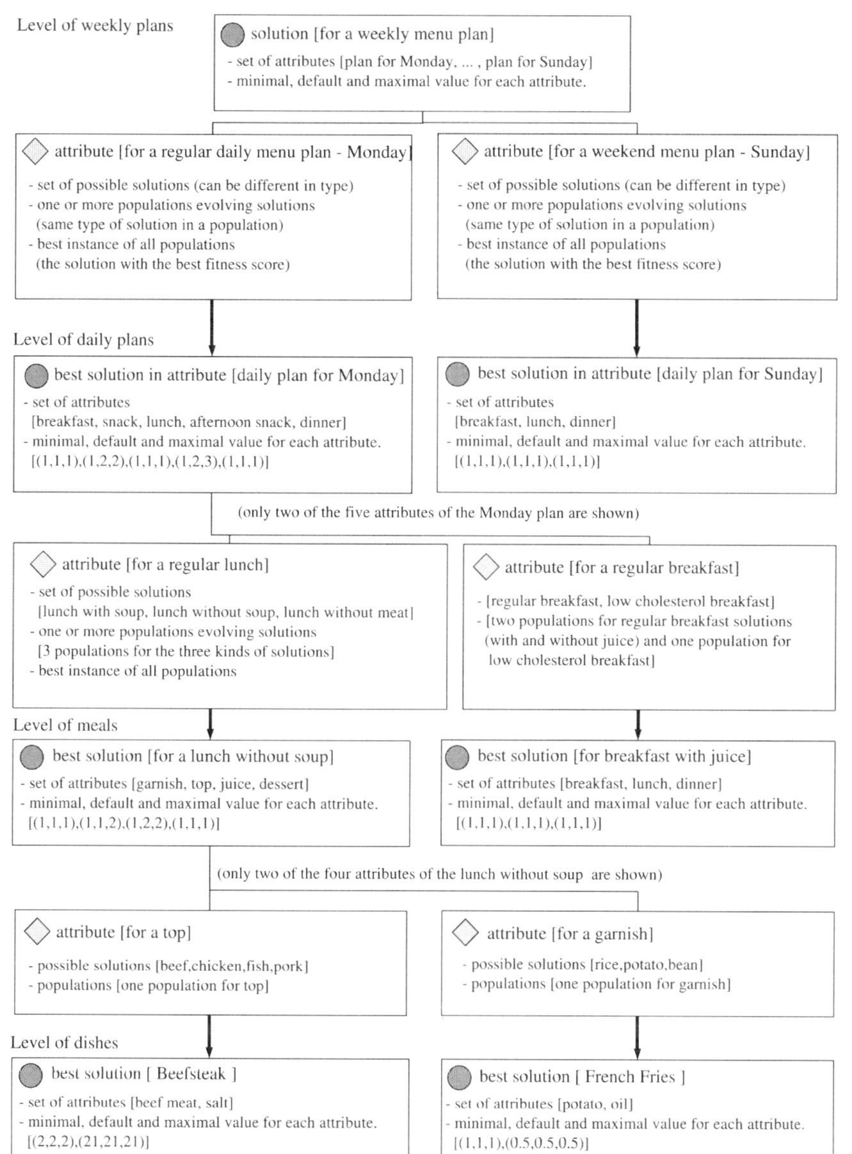

Fig. 3.2. The multi-level data structure of the algorithm. Concrete examples are in square brackets "[]"

our method can run the different level GAs in a synchronized manner. Table 3.2 gives an overview of the currently implemented mutation-based GA scheduling strategy and other possible alternatives as well.

Columns 4 and 5 of Table 3.2 show the sequence of how GSLib fires the evolution processes on the various levels for the top-down and bottom-up

Table 3.2. Various multi-level GA scheduling strategies (top-down, bottom-up, credit propagation, mutation based) at each algorithmic level

Level	Object	Number of instances	Top-Down			Bottom-Up			Credit Propagation	Mutation Based
Weekly	Population	1							Start with a defined amount of credit and decide on each level what to use it for	Fire evolution on lower levels
	Solution	40	1.	4.	7.	3.	6.	9.		when
	Attribute	280							1. for evolving the current level	
Daily	Solution	14000	2.	5.	8.	2.	5.	8.	2. use part of the credit and	mutation occurs
	Attribute	70000							share the other part among	
Meal	Solution	84e+05	3.	6.	9.	1.	4.	7.	the lower level objects	normal
	Attribute	336e+05							3. share all of the credit among	mutation
Dish	Solution	756e+07							the lower level objects	
	Attribute	6804e+07								
Nourishment	Solution	68e+09								
	Attribute	544e+09			no evolutionary process on these levels						
Nourishment Component	Solution	544e+09								

strategies, respectively. For example, the top-down strategy starts with an initial population (loaded randomly or from the case-base) and goes on by evolving the weekly level for a given number of iteration steps (multi-level iteration step: 1). Then, the evolution proceeds on the level of daily plans (2) by evolving the attributes of the weekly menu plans. After the second level, the process continues on the level of meals (3), and after that, the evolution restarts from the weekly level (4). The multi-level scheduling for credit propagation and mutation based strategy are also described in Table 3.2. In any strategy, there is no evolution on the level of dishes, nourishments, and nourishment components. The estimated number of instances of the solution and attribute classes is shown in column 3. Note that by exploiting a copy-on-write technique, most instances of the classes are virtual and stored in the same memory location.

In contrast to the sequential method, we keep all of the adjustable parameters of the various levels in the memory to provide a larger search space in which generally better solutions can be found. So, if a GA evolving daily menu plans can't satisfy its constraints and rules, its fact base (which consists of meals) can be improved by further evolving the populations on the level of meals.

It is also possible to generate partial nutritional plans. If one or more parameters have been set previously, the unassigned ones can be generated in a way that the whole plan satisfies the relevant constraints. A practical application of this feature is when one eats at his/her workplace and can't choose his/her lunch for the weekdays. In this case, the lunches of the weekly plan can be defined by the end user at the beginning of the week and MenuGene can develop the whole weekly menu plan without changing these lunches. The algorithm is capable to adjust the weekly plan to compensate for deviations. From a quantitative point of view, it is more important to keep the nutritional constraints on a weekly basis, than to keep them on a daily basis or in a meal. For this reason, MenuGene allows relatively more deviation from the optimal values on the lower levels (day/meal), but it tries to stick to the constraints on a weekly basis.

3.6 Ensuring Harmony

Even for the calculation of personalized numerical constraints on nutrients, dietetian expert knowledge may be needed. This sort of knowledge is, however, much easier to formalize than the ability to assess harmony. As noted in Sect. 3.2, harmony has several aspects like taste, color, religion, season of the year etc. The knowledge of an expert could be roughly described as rules in these aspects involving sets or types of dishes. The second conceptual level could contain guidelines for dealing with conflicting rules.

In our test system, we tried to model expert knowledge on harmony by developing a domain ontology, i.e. a nested set structure, for the various aspects of harmony. For example, with respect to dominant taste, we have the sets of sour, sweet, bitter, salty and neutral sets. Then, all dishes are indexed by experts with respect to membership in the relevant sets. For example, French fries, as a dish, is a member of the dry garnish set, the set of dishes fried in oil, the set of dishes with light/brown color, the set of dishes with neutral taste and the set of starch based dishes. Rules on harmony are then defined for the co-occurrence of certain sets, and whenever we assess the harmony of a structure (e.g. daily menu) we check the set membership of its constituents. This set structure is quite similar to that of PRISM [11].

Harmony assessment is built in the fitness function. After the goodness is computed as a function of the numerical constraints (see Sect. 3.4.2), we examine which of the harmony rules are applicable. Each rule has two parts: conditions and fitness modification value. The general form of the rules is $r_i = \langle \text{condition}_1, \ldots, \text{condition}_n, \text{fitness modification value} \rangle$. The fitness value (which is less or equal to zero) should be divided by the modification value so that if it is less than one, the fitness will decrease.

$$\langle \text{meat} \rangle \rightarrow \langle \text{whitemeat} \rangle \mid \langle \text{redmeat} \rangle$$

$$\langle \text{white meat} \rangle \rightarrow \langle \text{chicken} \rangle \mid \langle \text{fish} \rangle , \quad \langle \text{red meat} \rangle \rightarrow \langle \text{beef} \rangle \mid \langle \text{pork} \rangle$$

$$\langle \text{lunch} \rangle \rightarrow \langle \text{lunch\_with\_white\_meat} \rangle \mid \langle \text{lunch\_with\_red\_meat} \rangle \mid \langle \text{vegetarian\_lunch} \rangle$$

The conditions part of the rules on the level of meals contains one or more dish sets (e.g. dry top) or specific dishes (e.g. tomato soup). Some example rules might look like these:

$$r_1 = \langle \text{dry top, dry garnish}, 0.75 \rangle , \quad r_2 = \langle \text{tomato soup, tomato drink}, 0.6 \rangle ,$$

$$r_3 = \langle \text{dry top, dry garnish, pickles}, 0.8 \rangle , \quad r_4 = \langle \text{candy}, 0.7 \rangle$$

Rule r_2 means that for each meal that contains tomato soup and tomato drink the system will replace the fitness with 60% of the original value. Rule r_3 penalizes the simultaneous occurrence of three dishes while rule r_4 penalizes solutions with any kind of candy.

Rules may be applied and configured for any level (e.g. daily level, meal level) of the algorithm. So, if the above rule r_2 is applied at the daily level, it will reduce the fitness to 60% of those daily menus that contain tomato soup and tomato drink anywhere in their meals.

Only the most appropriate rules are applied. For example, if we have a meal plan that contains tomato soup and tomato drink then from the rules, $r_2 = \langle \text{tomato soup, tomato drink}, 0.6 \rangle$ and $r_5 = \langle \text{soups, tomato drink}, 0.72 \rangle$ only the former is applied because it is more specific. Only the strictest rule is applied from two or more rules with the same condition parts.

Since MenuGene uses the rules as parameters, the rule-base of the system can be developed while the system is being used. Incremental development of the rule-base is similar to that implemented in MIKAS [13]. The increasing number of rules doesn't have an impact on the generation time of a plan, because the rule-base of the system is preprocessed.

According to experts, harmony is more important on lower levels. For example, a meal or a daily plan with two dishes or meals from tomato is not well assorted. Such plans can be excluded with simple rules.

3.7 Algorithmic Tests

In order to check the efficiency of the MenuGene system, we performed some tests using a commercially available food and nutrient database.

3.7.1 Convergence

Runtime performance is an important factor for MenuGene as it is planned to run as an on-line service. Runtime is determined by the number of operations to be performed in each generation. Not surprisingly, we observed a strong linear connection between the probabilities of mutation and crossover, and runtime. However, our main concern is the quality of the solution. So we examined the connection between the runtime and the quality of the solution in a wide range of algorithmic setups (adjusted parameters were number of iteration, population size, probability of mutation and crossover).

As Fig. 3.3 shows, although the quality of the solutions improves with longer runtimes (whether it is the effect of any of the adjusted parameters), the pace of improvement is very slow after a certain time, and, on the other hand, a solution of quite good quality is produced within this time. This means that it is enough to run the algorithm until the envelope curve starts to saturate.

The convergence of the algorithm was measured with test runs generating meal plans. Constraints were energy (min = 4190 kJ, opt = 4200 kJ, max = 4210 kJ) and protein (29 g, 29,27 g, 29,5 g), population size was 200,

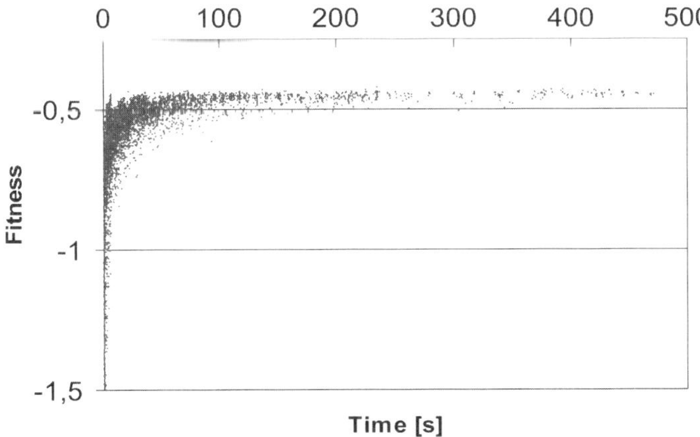

Fig. 3.3. Runtime of MenuGene in various algorithmic setups. Each dot represents the result of a test run

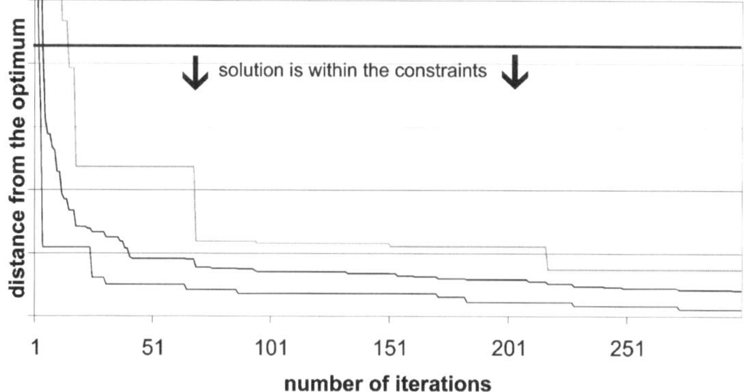

Fig. 3.4. Distance from the optimum for the worst (upper curve), average (middle curve) and best (lower curve) solution of 10 runs, in the function of the number of iterations

probability of crossover and mutation was 0.9 and 0.2. Results show that with two constraints, a satisfactory solution was found in 6.2 iterations on average (14 iterations in the worst case). As Fig. 3.4 shows, there is hardly any improvement in the quality of the solution after 250 iterations, so a nearly optimal plan can be found in this time.

3.7.2 Crossover and Mutation

We analyzed the effect of the crossover and mutation probabilities on the fitness. We used the same randomly generated initial populations for the tests, and averaged the results of ten runs in each configuration.

The results showed that while the probability of the crossover does not influence the fitness too much, a mutation rate well above 10% is desirable, particularly for smaller populations. This result is surprising at first, as the literature of GA generally does not recommend mutation rates above $0.1 \ldots 0.5\%$. However, due to the relatively large number of possible alleles, we need high mutation rates to ensure that all candidate alleles are actually considered in the evolution process.

3.7.3 The Effect of the Gradual Diminution of Constraints

We tested the reaction of the algorithm to the gradual diminution of constraints. Minimal and maximal values were two times the size of the suggested at the start and were gradually decreased to be virtually equal from the aspect of human nutrition. More than 150.000 tests were run.

The tests showed that our method is capable of generating nutritional plans, even where the minimal and maximal allowed values of one or two constraints are virtually equal and the algorithm finds a nearly optimal solution when there are three or four constraints of this kind. According to our nutritionist, there is no need for constraints with virtually equal minimal and maximal values, and in most pathological cases the strict regulation of four parameters is sufficient. Our method proved capable of generating menus with meal plans that satisfy all constraints for non-pathological nutrition.

3.7.4 Runtimes of the Multi-level Algorithm

Generated plans should satisfy numerical constraints on the level of meals, daily plans and weekly plans. The multi-level generation was tested with random and real world data. The tests showed that for a mainstream desktop personal computer it takes between ten and fifteen minutes to generate a weekly menu plan with a randomly initialized population. The weekly menus satisfied numerical constraints on the level of meals, daily plans and weekly plans. Our tests showed that the rule-based classification method successfully omits components that don't go well together.

The case-based initialization of the startup population increases the speed of the generation process. Whenever a solution is needed for a plan with constraints that has been made previously, it would be enough to use the solution that can be found in the case-base for these constraints. However, with some iteration, the algorithm may find better solutions than are in its initial population at startup. If there was no improvement in the best solution stored in the database for a particular plan, it can be assumed that one of the best solutions was found for that menu plan.

3.7.5 Variety of Successive Solutions

Variety is also an important factor considering dietary plans. GAs use random choice for guiding the evolution process for near-optimum search, so if the search space is large enough, the solution found should be close to the optimum, but need not be similar in several consecutive runs. However, GAs are known for finding near-optimal solutions, so if there are strict numerical constraints, then it can easily happen that only a small subset of solutions satisfies them (which are close enough to the optimum), and the probability that the solutions don't have similar attributes is marginal. So, a method for adjusting the expected occurrence of the alleles of the GA is needed for providing sufficient variety in the menu plans.

We measured the variety of menu plans with constraints for regular dietary plans for women aged between 19–31 with mental occupation. The variety of the allele that represents one from the 150 possible soups for a regular lunch is shown Fig. 3.5. The figure shows the occurrence (ordered by frequency) of each of the 120 soups that were present more than 15 times (0.1%) in the best solutions in 15.000 test runs. The most frequent soup in the best solutions of 15.000 runs was present 482 times ($\sim$3.2%), the second 426 times ($\sim$2.8%) and the 50^{th} 102 times ($\sim$0.7%). Figure 3.5 also shows (lower part) the goodness of the best solution to which the corresponding alleles belong. The goodness

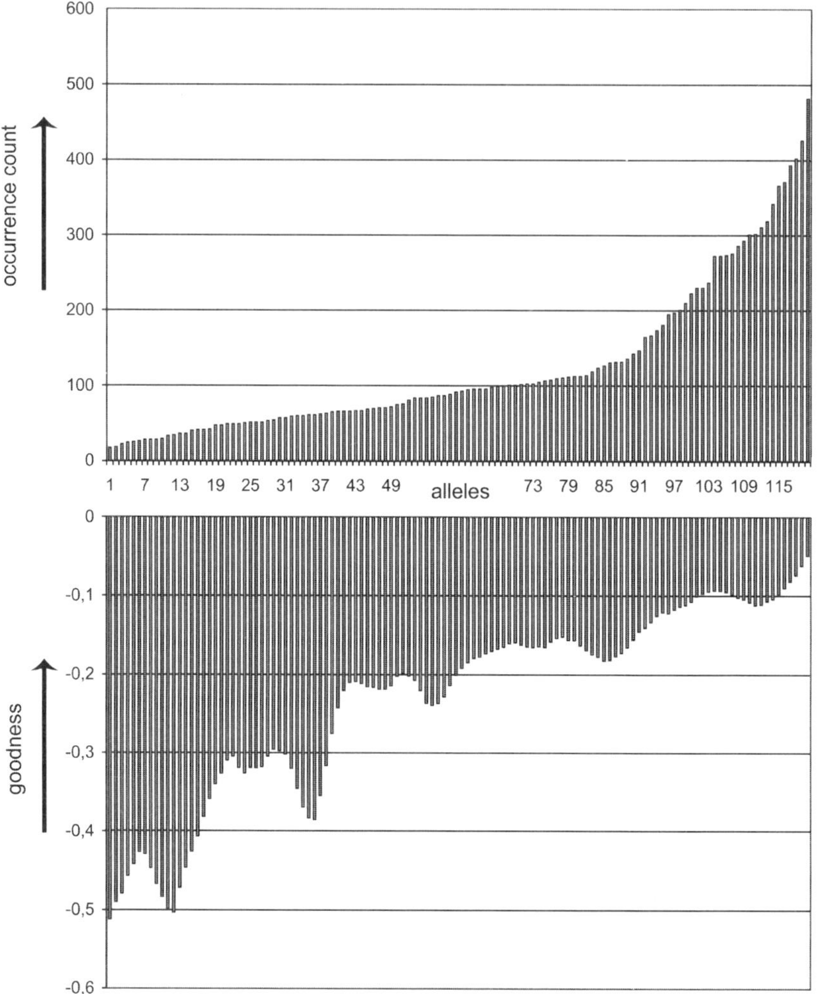

Fig. 3.5. The occurrences (ordered by relative frequency, upper figure) of 120 of the 150 possible alleles (soups) in the best solutions (lunches) of 15.000 runs. The lower figure shows the goodness of the best solution which the respective soup was part of

of an allele is computed by summing its best fitness (i.e. the best fitness value of all solutions the allele was part of) with the weighted best fitness values of its 8 neighbors. The goodness of the ith allele is defined as

$$g[i] = f[i] + \sum_{j=1}^{4} [(1 - 0.2j) \cdot (f[i - j] + f[i + j])]$$

where $f[i]$ is the fitness of the ith allele. The trend curve (lower part) shows that solutions containing frequent alleles have generally better goodness.

The results show, that alleles appearing in good solutions are used more often by the algorithm and the frequency of usage is about inverse proportional to the fitness of the best solution generated by using the particular allele.

However, it may happen that a run of the algorithm with properly configured nutritional constraints and rules results in a dietary plan that contains several occurrences of same dishes or dishes made from the same ingredients. Therefore, we allow more general rules, like $r_s = \langle ?, ?, 0.5 \rangle$ to be recorded in our rule-base, which also get pre-processed during the initialization of the algorithm. Rule r_s will penalize every solution that has the same value (solution) represented by its attributes more than one time. So, if r_s is imposed on a daily menu plan which contains orange drink for breakfast and lunch as well, then the fitness of this daily plan will be reduced by 50%.

We measured the effect of the rules on the variety and mean occurrence of the alleles (drinks) considering the solutions for meal plan (lunch). The results of the statistical analysis are shown in Table 3.3. Two rules (r_A,r_B) where imposed on two alleles, respectively. The strictness of the rules was decreased from 100% to 75%, 50% and finally to 25%, giving a total of 16 configurations. Rule r_A penalized the solutions that contained drink "A" while r_B penalized solutions with drink "B". The relative occurrences of "A" are shown in the function of the strictness of the rules in Fig. 3.6.

We employed two-sample Kolmogorov-Smirnov goodness-of-fit hypothesis test with the significance level of 5% to the two random samples created by recording the alleles representing drinks in neighboring configurations, running 1.000 times each (using 10 random populations, running each one 100 times). In the table lists those P values (denoted with KS) for which the Kolmogorov-Smirnov test has shown significant difference in the distribution of the two independent samples.

We observed significant difference for all of the test pairs with respect to increasing the strictness of r_A (rows in Table 3.3), however the same was not true for all of the pairs with respect to increasing the strictness of r_B (columns in Table 3.3), so P values are not listed for such pairs. The explanation of this phenomenon is hidden in the differences between the occurrences of alleles A and B without penalties, which are 436 (43.6%) for "A" and 68 (6.8%) for "B", out of the 1.000 possible. Since the number of instances of "A" is comparable to the possible instances, rule r_A not only changes the mean occurrence of "A", but significantly changes the distribution of the alleles. In case of penalizing the meals with drink "A" by 75%, the occurrence count of "A" decreases by 133 ($\sim$30%) from 436 (100%) to 303 ($\sim$70%). As Fig. 3.7 shows, the 103 occurrences are shared somewhat proportional among the other alleles ("A" is 15th, "B" is the 12th allele in Fig. 3.7).

We performed single sample Lilliefors hypothesis test of composite normality on the samples with an element size of 10 on 100 runs of the algorithm with 10 different starting populations and counting the occurrences of "A" and "B". The distribution of the occurrences of "A" in function of the starting populations proved normal, except for one case. Again, due to the few

Table 3.3. Statistical analysis of the distribution of the potential alleles (drinks) in the best solutions (lunches) and the mean occurrence of the alleles (A,B) on which the rules were imposed

		A													
		100%				75%				50%				25%	
				KS	3.1822e-004			KS	5.3952e-005			KS	3.1822e-004		
	100%	A	100%			A	69,50%			A	31,42%			A	7,80%
		B	100%	T	7.4482e-006	B	139,71%	T	4.5663e-007	B	175,00%	T	9.4052e-005	B	186,76%
				Sr	0.0020			Sr	0.0020			Sr	0.0020		
		Sr	0.0020			Sr	0.0020			Sr	0.0020			Sr	0.0020
				KS	2.1326e-007			KS	9.4868e-007			KS	0.0031		
	75%	A	103,44%			A	69,50%			A	38,53%			A	10,09%
		B	32,35%	T	6.4855e-006	B	32,35%	T	9.9226e-005	B	33,82%	T	3.8001e-006	B	41,18%
B				Sr	0.0020			Sr	0.0020			Sr	0.0020		
		Sr	–			Sr	0.0078			Sr	–			Sr	0.0059
				KS	3.1822e-004			KS	7.5721e-008			KS	5.8152e-007		
	50%	A	96,79%			A	71,33%			A	35,78%			A	9,63%
		B	17,65%	T	2.1538e-004	B	13,24%	T	2.0654e-007	B	17,65%	T	4.8058e-005	B	14,71%
				Sr	0.0039			Sr	0.0020			Sr	0.0020		
		Sr	0.0156			Sr	–			Sr	0.0313			Sr	0.0156
				KS	2.7488e-007			KS	1.2116e-005			KS	1.2116e-005		
	25%	A	113,76%			A	80,73%			A	35,55%			A	8,26%
		B	1,47%	T	3.4505e-004			T	−(*)			T	−(*)		
						B	5,88%			B	5,88%			B	4,41%
				Sr	0.0020			Sr	0.0020			Sr	0.0020		

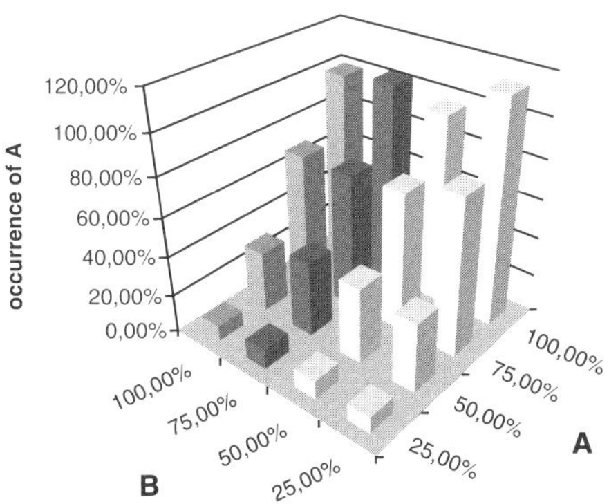

Fig. 3.6. The relative occurrence of a particular solution (A) in function of the strictness of two rules (penalizing solution A and B)

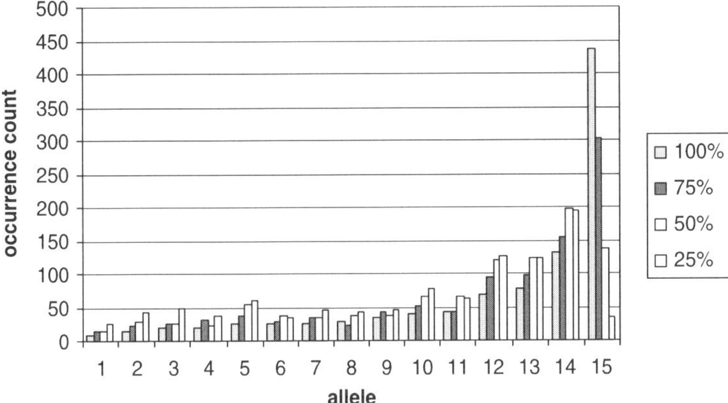

Fig. 3.7. Occurrence counts of the 15 possible alleles (drinks) in a solution (for lunch) in 1.000 runs in function of the strictness of the rule imposed on the 15th allele

occurrences of "B" in the test runs, we could not determine its distribution. We paired the samples of neighboring configurations and if both had normal distribution with a significance level of 5%, we employed paired t-tests to check, if there is a significant difference in the mean occurrences of alleles "A" and "B". The results of the paired t-tests are shown in Table 3.3, denoted with T. In case of significant differences, the corresponding P values are also listed. If one of the samples did not have normal distribution, we marked the case with an asterisk (*). Since the sample distribution was not known for more than half of the samples, we employed the Wilcoxon signed rank test of equality of medians with the significance level of 5% on each sample pair, to measure whether there is significant difference between the mean occurrences. Results are shown in Table 3.3 with the corresponding P values, and are denoted with *Sr*. There were only 3 situations where there was no significant difference between the mean occurrences. These cases are marked with a hyphen $(-)$.

3.8 Conclusions and Future Work

This chapter described the results of the automatic, parameterized menu planner MenuGene that uses multi-level, multi-objective Genetic Algorithms for a near-optimum search. Our fitness function classifies menus according to the amount of nutrients and harmony. From the aspect of nutritional constraints, our method outperforms present-day nutrition planning systems. Recently developed nutrition planning systems, such as CAMPER [9] maintain the constraints only on a daily basis, in contrast, MenuGene satisfies constraints on the meal-by-meal, daily and weekly levels.

We proposed an abstract scheme for the consistent handling of the different level problems, for the implementation of crossover and mutation and for the coding of chromosomes as well as a fitness function. Algorithmic tests revealed that a relatively high mutation rate is desirable. This can be explained by the fact that mutation is the only operator which can introduce new genetic information in the population. It was also shown that after a certain time, the quality of the solution does not improve much. Our tests showed that GAs generally produce high variety, at least in non-constrained configurations, but in any case, rules can be used to employ the desired level of variety and harmony.

The advantage of our approach is that it uses the same algorithm on every level, thus the hierarchical structure is easily expandable. The method is capable of controlling the nutrition on longer periods. Monthly optimizations could be performed, without the need to plan the whole monthly plan in one run. After the first week, the plan for the second week can be made with the previous weekly plan in mind.

Future work on MenuGene includes the development of parallel computation methods for improving the runtime of the co-evolution process and the continuous improvement of MenuGene's case-base and rule-base with a web-based application that was developed for human experts. An enhanced version of the rule system will also support time dependent rules to express locality in time.

In a more complex user interface, users could also define their personalized rules (e.g. to exclude some dishes). These rules would only be used for the user that defined them and have lower priority than the rules given by experts.

Acknowledgements

The work presented was supported by the National Research and Development Program #NKFP 2/052/2001 and the Hungarian Ministry of Health.

References

1. The interactive menu planner of the National Heart, Lung, and Blood Institute at http://hin.nhlbi.nih.gov/menuplanner/ [Verified June 2006]
2. The Cordelia Dietary and Lifestyle counseling project at http://cordelia.vein.hu/ [Verified June 2006]
3. Balintfy, J. L.: Menu Planning by Computer, Communications of the ACM, vol. 7, no. 4, pp. 255–259., April, 1964
4. Dollahite J, Franklin D, McNew R. Problems encountered in meeting the Recommended Dietary Allowances for menus designed according to the Dietary Guidelines for Americans. J Am Diet Assoc. 1995 Mar;95(3):341–4, 347; quiz 345–6
5. Food and Nutrition Board (FNB), Institute of Medicine (IOM): Dietary Reference Intakes: Applications in Dietary Planning, National Academy Press. Washington, DC. 2003
6. Food and Nutrition Board (FNB), Institute of Medicine (IOM): Dietary Reference Intakes for Energy, Carbohydrate, Fiber, Fat, Fatty Acids, Cholesterol, Protein, and Amino Acids (Macronutrients), National Academy Press. Washington, DC. 2002
7. Eckstein EF. Menu planning by computer: the random approach. J Am Diet Assoc 1967 Dec;51(6):529–533
8. Hinrichs, R. R. Problem Solving in Open Worlds: A Case Study in Design. Erlbaum, Northvale, NJ. 1992
9. C.R. Marling, G.J. Petot, L.S. Sterling Integrating Case-Based and Rule-Based Reasoning to Meet Multiple Design Constraints
10. Petot G.J., Marling C.R. and Sterling L. An artificial intelligence system for computer-assisted menu planning. Journal of the American Dietetic Association; 98: 1009–1014, 1998
11. Kovacic KJ. Using common-sense knowledge for computer menu planning [PhD dissertation]. Cleveland, Ohio: Case Western Reserve University; 1995
12. Khan AS, Hoffmann A. An advanced artificial intelligence tool for menu design.Nutr Health. 2003;17(1):43–53
13. Khan AS, Hoffmann A. Building a case-based diet recommendation system without a knowledge engineer.Artif Intell Med. 2003 Feb;27(2):155–79
14. Noah S, Abdullah S, Shahar S, Abdul-Hamid H, Khairudin N, Yusoff M, Ghazali R, Mohd-Yusoff N, Shafii N, Abdul-Manaf ZDietPal: A Web-Based Dietary Menu-Generating and Management System Journal of Medical Internet Research 2004;6(1):e4
15. Bucolo M, Fortuna L, Frasca M, La Rosa M, Virzi MC, Shannahoff-Khalsa D. A nonlinear circuit architecture for magnetoencephalographic signal analysis.Methods Inf Med. 2004;43(1):89–93
16. Laurikkala J, Juhola M, Lammi S, Viikki K. Comparison of genetic algorithms and other classification methods in the diagnosis of female urinary incontinence.Methods Inf Med. 1999 Jun;38(2):125–31
17. Carlos Andrés Pena-Reyes, Moshe Sipper Evolutionary computation in medicine: an overview, Artificial Intelligence in Medicine 19 (2000) 1–23
18. P.S. Heckerling, B.S. Gerber, T.G. Tape, R.S. Wigton Selection of Predictor Variables for Pneumonia Using Neural Networks and Genetic Algorithms, Methods Inf Med 2005; 44: 89–97

19. Coello Coello, C.A.: A comprehensive survey of evolutionary-based multi-objective optimization techniques, Int. J. Knowledge Inform. Syst 1, 269–309, 1999
20. Multi-level Multi-objective Genetic Algorithm Using Entropy to Preserve Diversity, EMO 2003, LNCS 2632, pp. 148–161, 2003
21. The M.I.T. GALib C + + Library of Genetic Algorithm Components at http://lancet.mit.edu/ga/ [Verified June 2006]

Evaluation of Healthcare IT Applications: The User Acceptance Perspective

Kai Zheng[1], Rema Padman[2], Michael P. Johnson[3], and Herbert S. Diamond[4]

[1] The University of Michigan kzheng@umich.edu
[2] Carnegie Mellon University rpadman@cmu.edu
[3] Carnegie Mellon University johnson2@andrew.cmu.edu
[4] The Western Pennsylvania Hospital hdiamond@wpahs.org

As healthcare costs continue to spiral upward, healthcare institutions are under enormous pressure to create cost efficient systems without risking quality of care. Healthcare IT applications provide considerable promises for achieving this multifaceted goal through managing inofrmation, reducing costs, and facilitating total quality management and continuous quality improvement programs. However, the desired outcome can not be achieved if these applications are not being used.

In order to better predict, explain, and increase the usage of IT, it is of vital importance to understand the antecedents of end users' IT adoption decisions. This chapter first reviews the theoretical background of *intention models* that have been widely used to study factors governing IT acceptance, with particular focus on the technology acceptance model (TAM)—a prevalent technology adoption theory in the area of information system research. Although TAM has been extensively tested and shown to be a robust, powerful, and parsimonious model, its limitations have also been recognized. The second part of this chapter analyzes these limitations and discusses possible precautions of potential pitfalls. The third part of this chapter specifically addresses the applicability of the technology acceptance model in the professional context of physicians, with a review of available studies that have applied TAM to the technology adoption issues in healthcare.

4.1 Introduction

Information systems pervade modern organizations, however, they cannot improve performance if they are not used (e.g., Davis et al., 1989). It has been widely acknowledged that individual acceptance of information technologies is a crucial factor in determining IT success. Although organizations make primary IT procurement decisions, the true value of an IT cannot appear until the end-users incorporate it into their work processes.

K. Zheng et al.: *Evaluation of Healthcare IT Applications: The User Acceptance Perspective*, Studies in Computational Intelligence (SCI) **65**, 49–78 (2007)
www.springerlink.com   © Springer-Verlag Berlin Heidelberg 2007

Resistance to end user systems by managers and professionals has been a widespread problem [33]. It is also well recognized that end users are often unwilling to use available systems that could generate significant performance gains [69, 80, 86]. Even in scenarios such as mandated use, i.e., use of an IT is required to perform a job, inadequate levels of individual acceptance can diminish the long-term value of the IT [3]. Under extreme situations it can lead to serious consequences [62].

In order to better predict, explain, and increase user acceptance of IT applications, we need to understand why people accept or reject an IT. Many factors may account for individuals' decisions in this regard: system design features (e.g. technical design characters), industry or workplace norms (e.g. professional autonomy), and individual differences (e.g. cognitive styles). Realizing the importance of user acceptance and the complex, elusive causal relations governing users' adoption decisions, a considerable stream of information system research has attempted to theorize and explain the antecedents of user acceptance of IT. The main questions that IS researchers want to answer are: 1) while it is evident that individuals demonstrate different behavior towards embracing a new information technology, what causes these differences? 2) how can we theorize these factors into a parsimonious list of variables that are important, consistent antecedents of individual IT acceptance; 3) what tactics and corrective actions can be used by managers and developers to alleviate adoption variation; in particular, to eliminate problems associated with rejection of an IT? With a deeper understanding of these questions, researchers and practitioners can better predict how end users would respond to an IT in the early stages of system development or implementation process, which provides promise for improved user acceptance and therefore increased rate of IT success.

The majority of this stream of IS research draws upon well established, widely validated theories from social psychology. Notably, four theories have gained popularity and have been represented in various IS variations to study IT acceptance issues specifically. These theories are: theory of reasoned action (TRA, 1967), theory of planned behavior (TPB, 1988), diffusion of innovations (DOI, 1983), and social cognitive theory (SCT, 1963). A number of native IS theories, by and large influenced by the social psychology theories, has also been developed, such as the model of PC utilization (MPCU, 1991) and task technology fit (TTF, 1995). While each of these theories has considerable number of proponents in the IS domain, theory of reasoned action (TRA), theory of planned behavior (TPB), and TRA's IS adaptation, the technology acceptance model (TAM, 1986), have received most attention. In this chapter, we will mainly focus on the streams of IS research based on TRA, TPB, and TAM.

The rest of this chapter is organized as follows. Section 4.2 reviews the theoretical background of *intention models* that have been widely used in studying user acceptance of information technologies. In particular, Sect. 4.2.2 elaborates on the history, applications, and extensions of a prevalent theory—the technology acceptance model (TAM). Following the introduction to the existing models, Sect. 4.3 discusses their known limitations and Sect. 4.4 addresses specific issues associated with applying these models in the professional context of physicians. Section 4.4 reviews previous TAM or TAM-based studies on physicians' adoption and utilization of health IT applications. The last section presents some concluding remarks.

4.2 Theoretical Background

4.2.1 Social Psychology Theories

The theory of reasoned action (TRA) and theory of planned behavior (TPB) are widely studied theories of social psychology that address the determinants of consciously intended **behavior (B)**. Both theories are *intention models* which posit that the fundamental determinant of a person's consciously intended behaviors is **behavioral intention (BI)**.

According to TRA, behavioral intention is an additive combination of **attitude (A)** toward performing the behavior and **subjective norm (SN)** [7, 40]. "Attitude" refers to individual's positive or negative feelings about performing the focal behavior in question, which is determined by his or her salient beliefs (b_i) about consequences of performing the behavior multiplied by the evaluation of these consequences (e_i). "Subjective norm" captures a person's perception that most people who are important to him or her think the person should or should not perform the target behavior. The perception is determined by a multiplicative function of his or her normative beliefs (n_{bi}), i.e., whether perceived expectations of specific referent individuals or groups, and his or her motivation to comply with these expectations (m_{ci}). Mathematical expressions for these three constructs follow:

$$BI = A + SN \tag{4.1}$$

$$A = \sum b_i e_i \tag{4.2}$$

$$SN = \sum nb_i mc_i \tag{4.3}$$

While TRA works most successfully when applied to behaviors that are under a person's volitional control, it is unable to fully represent situations in which intervening environmental conditions are in place, such as limited ability, time, and resources. This motivates the development of an extension of TRA called the theory of planned behavior (TPB). TPB adds a third determinant of behavior intention, **perceived behavioral control (PBC)**, to the TRA framework [5,6]. Perceived behavioral control indicates a person's perception concerning how difficult the behaviors are and how successfully the individual can, or can not, perform the activity. PBC is postulated to have a direct impact on behavior; it also influences the target behavior indirectly via

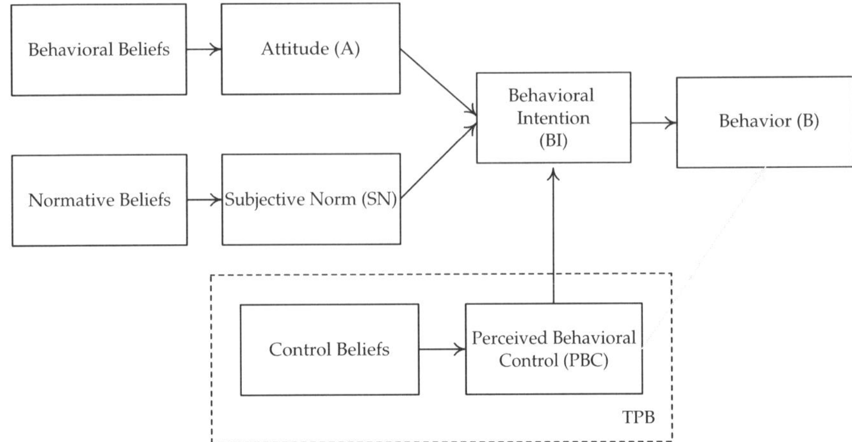

Fig. 4.1. Theory of Reasoned Action and Theory of Planned Behavior

behavioral intention. PBC is determined by the total set of accessible control beliefs (c_i), i.e., beliefs about the presence of factors that may facilitate or impede performance of the behavior, weighted by the perceived power (p_i) of the control factor:

$$BI = A + SN + PBC \qquad (4.4)$$

$$PBC = \sum c_i p_i \qquad (4.5)$$

Figure 4.1 illustrates the constructs and their relations as posited by the theories of reasoned action and planned behavior. These two theories have been extensively tested in empirical studies spanning a wide variety of subject areas, such as dishonest actions [16], driving violations [72], condom use [8,55], physician activities [43,46], smoking cessation [29,45], and substance use [68]. Meta analyses have shown that their predictive powers are substantially supported [11,79].

4.2.2 The Technology Acceptance Model

The most well-known IT acceptance theory in IS, the technology acceptance model (TAM), is an adaptation of TRA that is specifically designed to study user acceptation of computer systems. The goal of TAM is to "provide an explanation of the determinants of computer acceptance that is general, capable of explaining user behavior across a broad range of end user computing technologies and user populations, while at the same time being both parsimonious and theoretically justified" (Davis et al. 1989, page 985). As Szajna (1996) indicated, "TAM is intended to resolve the previous mixed and inconclusive research findings associating various beliefs and attitudes with IS acceptance. It has the potential to integrate various development, implementation, and usage research streams in IS".

The theoretical foundation of TRA is the assumption that behavioral intention influences actual behavior. Davis (1986) used this insight to propose that IT acceptance behavior, **actual system use (U)**, is determined by a person's **behavioral intention to use (BI)**; this intention, in turn, is determined by the person's **attitudes towards using (A)** and his or her **perceived usefulness (PU)** of the IT. In TAM, attitudes towards use are formed from two beliefs: **perceived usefulness (PU)** of the IT and its **perceived ease of use (PEoU)**. All external variables, such as system design characteristics, user characteristics, task characteristics, nature of the development or implementation process, political influences, organization structure and so on, are expected to influence acceptance behavior indirectly by affecting beliefs, attitudes, and intentions [32].

$$BI = A + PU \tag{4.6}$$

$$A = PU + PEoU \tag{4.7}$$

$$PU = PEoU + External\ Variables \tag{4.8}$$

PU and PEoU are two fundamental determinants of TAM. Perceived usefulness (PU) is defined as "the degree to which a person believes that using a particular system would enhance his or her job performance", whereas perceived ease of use (PEoU) refers to "the degree to which a person believes that using a particular system would be free of effort". Figure 4.2 depicts these basic constructs and their relations.

The TAM model was empirically tested in a study of acceptance of a word processor among MBA students [33]. The study employed a 14-week longitudinal design, with two questionnaires administered, one after a brief introduction to the system and one at end of the 14-week period; self-reported frequency of use of the system every week was used to surrogate the actual usage behavior. The results only partially supported the model's constructs and relations that hold among them: attitudes were found to only partially mediate PU and PEoU; and the influence of PEoU on BI fluctuates depending on the phrase of use. That is, in the pre-implementation stage, PEoU directly influences BI, while in the post-implementation phase, its influence is fully mediated

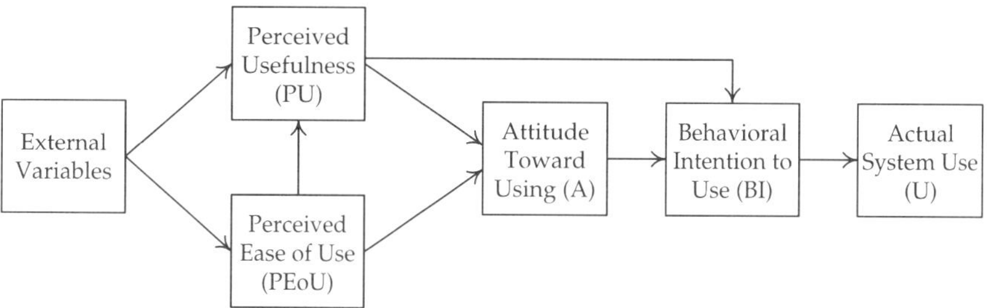

Fig. 4.2. Original Technology Acceptance Model

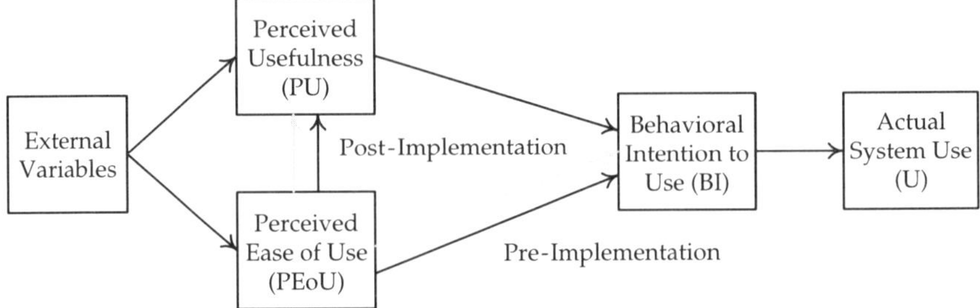

Fig. 4.3. Revised Technology Acceptance Model

via PU. As a result, Davis et al. (1989) suggested a revision of the original model that contains only three theoretical constructs: BI, PU, and PEoU. In addition, PEoU is postulated in post-implementation as a causal antecedent to PU, as opposed to a parallel, direct determination of BI. Figure 4.3 depicts the constructs and their relations in the revised TAM model.

TAM deviates from TRA in three important aspects. First, subjective norm, as one of the determinants of behavioral intention in TRA, was dropped from TAM because of its uncertain theoretical and psychometric status (Davis et al., 1989 page 986)[5]. Second, PU, as one of the salient beliefs that form a person's attitude, is postulated to directly influence behavioral intention. Third, the revised TAM model dropped the construct of attitudes towards use, which is a fundamental construct within TRA that mediates the influences of beliefs on behavioral intention. Compared to theory of planned behavior (TPB), TAM does not include perceived behavioral control (PBC), that is, the beliefs about the presence of factors that may facilitate or impede performance of the behavior.

Applications of TAM

Since its introduction in 1989, the technology acceptance model has been embraced by the IS community mainly for its two merits: 1) power in predicting IT acceptance; and 2) simplicity and ease of use due to its parsimonious constructs. As of 2003, the *Social Science Citation Index* listed 698 journal citations to the two articles that introduced TAM, i.e., Davis 1989, Davis et al. 1989 [58].

In these confirmatory, empirical studies, TAM has been applied to examine acceptance behavior across a wide variety of information technologies, user populations, and implementation contexts. For example Adams et al. (1992) applied TAM to study five different applications, namely electronic mail, voice mail, word processing, spreadsheets, and graphics. Davis (1993) replicated his

---

[5] Subjective norm is included again in the second version of TAM, as one of the determinants in social influence processes.

original study to examine use of electronic mail software and text editors, Sambamurthy and Chin (1994) used TAM in studying group decision support systems, and Subramanian (1994) applied TAM to study acceptance behavior of two mailing systems. Another stream of research compared TAM with alternative models. Davis et al (1989) reported that TAM outperforms its ancestor, TRA, in its predictive power of IT acceptance. Compared with the theory of planned behavior (TPB), TAM is also found to offer a slight empirical advantage; in addition, it is a much simpler, easier to use model to explain users' technology acceptance (Mathieson 1991, Hubona and Cheney 1994). Finally, Tayler and Todd (1995) compared TAM, TBP, and Decomposed TBP (DTBP)[6], and found that while TBP and DTPB increased the explained variance of use intention to 5% and 8%, respectively, they paid a high cost by adding more number of variables.

These studies consistently report that TAM explains a substantial proportion of the variance in usage intentions and behavior, typically around 40%[7], and PU is a strong determinant of behavioral intentions, coefficients are typically around 0.6 [95]. For a detailed list of these studies and the assembled view of their findings, please see the meta-analysis conducted by Lee et al. (2003).

In summary, TAM has been considered as a robust, powerful, parsimonious theory that is particularly useful in explaining and predicting IT acceptance behavior.

Evolution of TAM

As more confirmatory, empirical studies were conducted, certain limitations of TAM began to emerge. A fundamental assumption of TAM is that PEoU and PU, viewed as internal psychological variables, fully mediate the influence of uncontrollable environmental variables as well as controllable interventions on user behavior (external variables, as shown in Fig. 4.2 and 4.3). This assumption is based on TRA's assertion that any other factors that influence behavior do so only indirectly by influencing attitude and social norm, or their relative weights. This aspect of TRA has helped TAM achieve a set of parsimonious constructs that explain substantive behavioral variation; however researchers consistently reported that 1) a closer examination on these factors can help

---

[6] Decomposed TBP posits that 1) perceived usefulness, perceived ease of use, and compatibility are antecedents of attitude; 2) peer influence and superiors' influence are antecedents of subjective norm; and 3) self-efficacy, resource-facilitating conditions, and technology facilitating conditions as determinants of perceived behavioral control.

[7] It is arguable whether 40% explanatory power is sufficient in explaining subtle user acceptance behavior; however we view it as high considering TAM's parsimonious constructs. TAM's successive models, such TAM2 and UTAUT, increased the explanatory power to about 60%, with a cost of an excessive number of additional variables.

better understand subtle individual IT acceptance behavior; and 2) under certain circumstances (e.g., voluntary use versus mandatory use), influences of such factors may not be fully mediated via PU and PEoU; sometimes they may have direct influence on behavioral intention over and above that ⟨ PU and PEoU. As a result, researchers have constantly called for inclusion of these external or moderating factors into the framework of TAM.

Theoretical Extensions

A number of studies were therefore conducted to identify antecedents of the major TAM construct, i.e., PU and PEoU. For example, Hartwick and Barki (1994) found that users' participation and involvement in system development influences their behavioral intention and system use; nevertheless, the influence is only important for voluntary users of a system[8]. Igbaria et al. (1995) found that training, computing support, and managerial support affect both PU and PEoU, as well as the ultimate usage behavior. Agarwal and Prasad (1999) studied five individual differences (role with regard to technology, tenure in workforce, level of education, prior/similar experiences, and participation in training); they found that participation in training is an antecedent of PU, and prior experiences, role with regard to technology, tenure in workplace, level of education can be used to predict PEoU. Finally Venkatesh and Davis (1996) and Venkatesh (2000) examined antecedents of PEoU. They found that variables such as anchor (computer self-efficacy, perceptions of external control, computer anxiety, and computer playfulness) and adjustments (perceived enjoyment and objective usability) are determinants of a person's perceived ease of use of an information technology.

Besides this stream of research, many studies have examined the influence of a variety of external or moderating factors on technology acceptance. Such factors include system quality [51], training [50], culture [81], gender [44], personal innovativeness [3], computer playfulness [2], social influences [96], and perceived user resources [64]. As a meta-analysis study reveals, three facets of moderating factors are consistently reported to be significantly influential on acceptance behavior [85]. These factors are: 1) organizational factors: voluntariness of IT use and nature of task and profession; 2) technology factors: technology complexity, purpose of using IT (work-oriented vs. entertainment-oriented), and individual vs. group technologies; and 3) individual factors: gender, intellectual capacity, experience, and age [85]. These studies have provided valuable extensions to TAM by investigating the boundary conditions of TAM, and these studies typically reported higher predictive and explanatory power using revised models. However, these additional factors are usually applicable only to a specific information technology or a specific user population, so the results may not be generalizable. In addition, some of these

---

[8] While Hartwick and Barki's study used TRA as the basic theoretical framework, they discussed the relations of user participation and involvement on TAM's constructs; so we include the study here.

studies did not provide explicit evidence whether these newly introduced variables directly influence behavioral intention and use behavior, or instead their influence can be seen to be mediated by the basic TAM constructs, i.e., PU and PEoU. Therefore caution needs to be used when introducing additional factors in studying information systems in a given context.

TAM2

Synthesizing a large volume of the prior empirical work, Venkatesh and Davis (2000) proposed a newer version of TAM, known as TAM2, to address the original model's inadequacies. The main motivation of TAM2 is "to extend TAM to include additional key determinants of TAM's perceived usefulness and usage intention constructs, and to understand how the effects of these determinants change with increasing user experience over time with the target system" [95].

TAM2 provides two additional theoretical constructs: social influence processes (subjective norm), consisting of compliance-driven subjective norm, internalization-driven subjective norm[9], and image (identification); and cognitive instrumental processes, consisting of job relevance, output quality, result demonstrability[10], and perceived ease of use. These factors have been consistently reported to have considerable influence on PU in previous empirical studies. In addition, two moderating factors are introduced: 1) voluntariness, which is postulated as a moderating factor on SN → BI; and 2) experience, which is included to examine how the effects of determinants change with increasing user experience over time.

Venkatesh and Davis (2000) tested the new model using a longitudinal study on four different systems at four organizations, two involving voluntary usage and two involving mandatory usage. Model constructs were operationalized using questionnaires administered at three points: pre-implementation (month 1), post-implementation (month 2 and month 3). Usage behavior is measured as self-reported length of time spent on use of the system everyday. Besides the well studied relationships between BI → B, PU → BI, and PEoU → BI, both social influence processes and cognitive instrumental processes were found to significantly influence user acceptance. Except output quality is found to interact with job relevance, all other additional constructs are found to have direct significant influence on PU; in addition SN → PU relationship was found to be significantly moderated by experience. The revised model is depicted in Fig. 4.4.

Noteworthily, TAM2 only adds one construct, compliance-driven SN, which directly influences BI under mandatory settings and when experience

---

[9] Internalization refers to "the process by which, when one perceives that an important referent thinks one should use a system, one incorporates the referent's belief into one's own belief structure" [56].

[10] Result demonstrability refers to "tangibility of the results of using the innovation" [67].

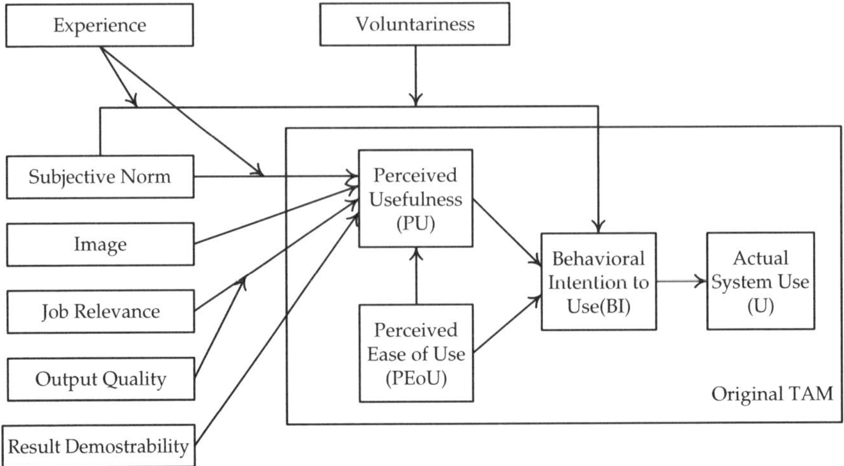

Fig. 4.4. Technology Acceptance Model 2

is in the early stages. As shown in Fig. 4.4, effects of other variables such as internalization-driven SN, image, and job relevance, are found to be fully mediated by PU.

Unified Theory of Acceptance and Use of Technology

Building upon a number of well established social psychology theories and their IS versions, the Unified Theory of Acceptance and Use of Technology (UTAUT) sets an ambitious goal to integrate the fragmented theories into a unified theoretical model. UTAUT draws on eight component theories: theory of reasoned action, technology acceptance model, motivational model, theory of planned behavior, a combined theory of planned behavior and technology acceptance model, model of PC utilization [89,90], innovation diffusion theory [75], and social cognitive theory [13,14].

By reviewing and consolidating these theories UTUAT posits that four key constructs: performance expectancy, effort expectancy, social influence, and facilitating conditions, are main determinants of behavioral intention and behavior. Four variables, including gender, age, experience, and voluntariness, are introduced to moderate the influences of these constructs. Figure 4.5 depicts these theoretical constructs and their relations.

UTAUT was empirically tested by Venkatesh et al. (2003). The results showed that the unified model outperforms each of the eight original models. The new models' theoretical constructs were found to account for 70% of the variance in usage intention [97]. Nevertheless UTAUT is relatively new and its validity needs to be substantially tested in more empirical studies. Compared to the original TAM framework UTAUT achieves 10% more predictive

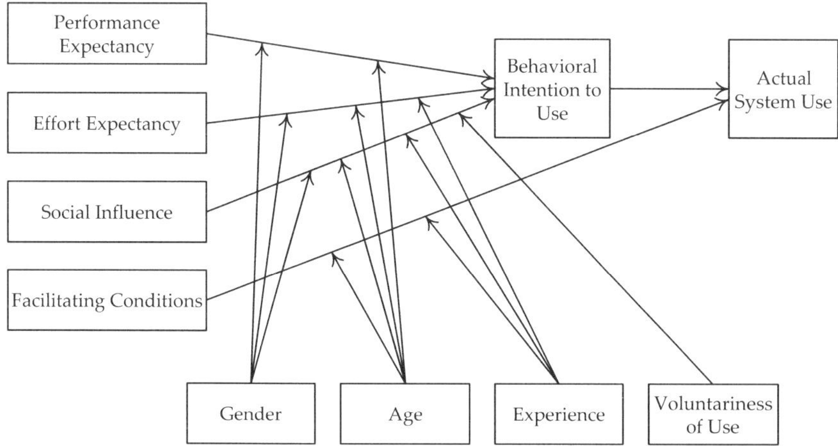

Fig. 4.5. Unified Theory of Acceptance and Use of Technology

power by replacing TAM's two constructs with eight variables. However the operationalization of the model constructs is therefore more difficult.

4.3 Limitation of the Existing Models

This section discusses limitations of TAM and other prevalent models. Some are well recognized issues that have received close and continued attention; some are emerging ones when new generation of information technologies are adopted. These new technologies (e.g., advanced decision support systems) usually represent more complex characteristics, require enhanced intellectual capacity by their end users, and may introduce novel role of computing that is subversive to the established, traditional work style and organization structure. Associated new IT acceptance behavior, therefore, may be beyond the capability of TAM to model or explain.

The present study mainly concerns physicians' IT acceptance behavior. Physicians are sophisticated professionals with high professional autonomy and control; the healthcare industry also has unique characteristics, such as vague definition of physicians' job performance, payers' role in care providers' motivation in adopting IT, etc. Therefore the well studied constructs listed above and their relations of TAM and TAM-like approaches may not be well suited to study physicians' adoption of health informatics applications. An in-depth discussion on this topic is presented in the next section.

4.3.1 Model Operationalization

PU, PEoU, attitude, BI, and use behavior are basic theoretical constructs defined in TAM. Operationalization of these constructs varies greatly from technology to technology, and is usually pertinent to the types of organizations and end users targeted by the technology. Unfortunately, most empirical studies have provided little justification of measures employed in operationalizing these constructs. This problem is particularly pronounced for use behavior, that is, actual usage of the technology in question. For example, to study adoption of an electronic mail application, choice of usage measures, such as frequency of use, time spent on reading and composing, or amount of messages sent, needs careful examination. In the domain of decision support systems, the problem is more severe: is frequency of use is the best measure of usage, or rather proportion of its decision-supporting advisories that receives user response? Improper choice of measure metrics may impede understanding of acceptance behavior and might lead to incorrect conclusions[11].

Concerned about detracting from the validity and reliability of TAM's survey instrument, many empirical studies used the original questionnaire items with only minor rewording, instead of evaluating their appropriateness in a specific context. For example, the statement "using this technology in my job would enable me to accomplish tasks more quickly"[12] is either irrelevant or prone to cause confusion for technologies that are not designed to directly improve effectiveness. The statement "I would find the technology easy to use" may result in divergent interpretations for sophisticated decision-supporting tools (e.g., financial prediction software and medical diagnosis systems), of which "operation" of these tools per se is straightforward but interpreting the produced results requires high-level intellectual capacity from end users. Adoption of instruments validated in distinct domains, therefore, needs to use caution. Even if an instrument appears to be valid, the pragmatic meaning of each question also needs close reexamination, beyond minor rewording to tailor the questions to a specific context.

Another problem in operationalizing TAM and its variants is that many of the empirical studies appear to focus on validating the theories, instead of investigating in contextual implications of the specific research task at hand. "It *(TAM)* has received disproportional amount of attention in IS research detracting research from more relevant research problems which may not be as easy to investigate rigorously" (commented by Alan Dennis in Lee et al. 2003, italics added). Researchers usually focused on measuring PU, PEoU, BI and analyzing their relations, ignoring other fundamental factors that actually drive the observed behavior. For example how much a person uses a particular word processor may depend less on the person's will than the volume of documents to be processed, and whether alternative applications are available. For computer-mediated communication technologies, such as electronic mail, usage may depend solely on size and activity of a person's social network

---

[11] Many studies, in fact, avoided this problem by measuring solely the behavioral intention, assuming actual behavior would be a natural consequence of behavioral intention. This introduces another dimension of problems, discussed in Sect. 4.3.5.

[12] This question and the next question were both adapted from Davis (1989).

(e.g., how many message to reply depends on how many messages he or she has received). In contrast, a person's perception of usefulness and ease of use of the technology or a particular email system may play a minor role. Many of the previous studies, however, focused too much on fitting TAM into a given context: if the model did not fit well, new constructs were introduced, instead of looking into the problem from new perspectives.

4.3.2 Self-Reported Measures

A notable limitation of TAM-based studies is that actual usage, i.e., usage behavior driven by intent to use, is usually surrogated by a person's self-reported usage. Typical questionnaire items assessing usage are "On average, how much time do you spend on the system every day? ______ hours and ______ minutes" and "How many times do you believe you use this system during a week?"[13]. While Venkatesh and Davis (2000) argued other research suggests self-report usage are more appropriate as relative measures [17, 47, 88] and the design of their empirical study avoided common method variance[14] by measuring self-reported usage and its determinants across time in different questionnaires, they acknowledged that researchers have also reported that self-reported usage measures are biased (Venkatesh and Davis, 2000, page 194). For example, Straub et al. (1995) found little similarity between self reported measures and computer recorded measures of IT usage. The between-method correlation was assessed as 0.293 and the use of these two sets of measures within a nomological context consisting of TAM's independent variables resulted in different path estimates, that is, self-reported measures of system usage are strongly related to self-reported PU and PEoU, while computer-recorded usage measures are not. As a result, researchers believe that the TAM model may need substantial reformulation or that IT usage needs to be divided between computer-recorded and self-reported usage [26].

Given practical impediments in accessing objective usage data (e.g., prohibitive cost or privacy concerns), self-reported measures appear the only means to approximate actual behavior. Unfortunately, a large body of research indicates that self-reports can be a highly unreliable source of data [77], and differences between perceived measures and objective measures are well acknowledged by psychologists and survey methodologists. These differences can be introduced by psychological processes such as positive illusions and cognitive consistency, or introduced by common method variance. For example, survey questions like "How many times do you believe you use this technology during a week" carries the risk of over-reporting (e.g., using a technology superficially in many sessions) as well as underreporting (e.g., using a technology extensively in each session while with less frequency); when use of an

---

[13] Adapted from Venkatesh and Davis (2000) and Malhotra and Galletta (1999), respectively.

[14] Common method variance refers to the variance that is attributable to the measurement method rather than the constructs the measures represent.

technology is frequent, respondents are also unlikely to have detailed representations of numerous individual episodes of use stored in memory. Questions such as "How many hours do you believe you use this system every week?" also make an implicit demand for respondents to remember and enumerate specific autobiographical episodes, which may in turn fail to solicit an accurate estimation [19]. Some of the TAM-based empirical studies employed longitudinal design, in which self-reported usage was measured at discrete time points (months 1, 2, and 3, for instance). As studies of cognitive processes indicate, respondents consistently tend to overestimate the number of events when a reference period includes a period more distant in time, referred to as telescoping; evidence has shown that overestimation due to telescoping effect may be as high as 32% [18].

Besides the self-reported measures of technology usage, other self-reported measures of perceptions (PU, PEoU, etc) can also be profoundly influenced by question wording, format, and context [77]. For example the halo effect[15] on TAM's contemporaneously measured constructs (PU and PEoU, for instance) may not be overlooked [33]; influences of social-desirability and self-presentation[16] can further contaminate self-reported data [36].

Although not examined before, Hawthorne effect may also exist in TAM-based empirical studies. Hawthorne effect is defined as "an experimental effect in the direction expected but not for the reason expected; i.e. a significant positive effect that turns out to have no causal basis in the theoretical motivation for the intervention, but is apparently due to the effect on the participants of knowing themselves to be studied in connection with the outcomes measured" [66]. It refers to the phenomena that certain behaviors are observed not because of intended interventions but because subjects are aware that they are being studied. In the context of IT acceptance research, individuals may alter their perceptions of an IT and subsequent use behavior (likely leaning upwards) due to such effect. In some TAM empirical studies perceptions and use behavior may also be altered due to the credibility and bases of researchers who introduced the technology; for example in the validation study in Davis et al. 1989, MBA students may perceive and use the word processor differently if it were not introduced by the researchers.

---

[15] "Halo effect" refers to the extension of an overall impression of a person (or one particular outstanding trait) to influence the total judgment of that person. The effect is to evaluate an individual high on many traits because of a belief that the individual is high on one trait. Similar to this is the "devil effect", whereby a person evaluates another as low on many traits because of a belief that the individual is low on one trait which is assumed to be critical.

[16] For example when reporting "learning to operate a technology would be easy for me", respondents with higher professional status or ranks may have a tendency to report it easy to learn, compared to respondents with lower status or ranks.

4.3.3 Type of Information Systems Studied

TAM has been applied to study acceptance behavior of a wide range of technologies, however, relatively few have studied complex information systems. Here, "complex information systems" refers to information systems whose usefulness and usability may not be immediately perceivable to its end users, while its true value would only derive from its long-term, sustainable use as an integral part of end users' work process. Examples of such complex information systems include sophisticated decision support systems, data analysis tools, and business intelligent systems. The majority of the healthcare applications such as computerized order entry systems and electronic medical record systems also fall into this category.

In the previous confirmatory, empirical research of TAM, information systems studied include word processing or spreadsheets packages, electronic mail, and operation systems. These are simple or popular technologies and their usefulness and ease of use can be readily understood; in addition, these technologies usually do not introduce "dramatic" impacts on characteristics of users' role, job, or tasks. However, usefulness and ease of use may not be so obvious for complex information technologies. These complex technologies are often associated with other consequences that may not be foreseen by either developers or end users, for example, interference with established workflow, detriment to professional autonomy, and shift of organizational structure. Theories and methods developed for the earlier generation of information systems therefore deserve a close examination for their applicability, efficiency, and effectiveness in studying new technologies [25]. Longitudinal studies, alternative approaches to TAM and its variants, and novel evaluation techniques are constantly called for in order to accommodate new issues brought up by complex information systems to ensure their longevity and return of investment.

4.3.4 Temporal Dynamics

Even though the initial decision regarding adaptation or rejection of an IT is an indispensable prerequisite, it may not be sufficient to lead to the IT's continued use. Factors that affect initial adoption may have no effect, or even the opposite effect on the later decisions to continue using the technology [91]. For example Karahanna et al. (1999) conducted a between-subjects comparison to study the impact of innovation characteristics on adoption (initial decision with no or little experience) and usage behavior (continued usage with greater experience). The results showed that for adoption, the significant predictors are relative advantage, ease of use, trialability, results demonstrability, and visibility. In contrast, for usage, only relative advantage and image are significant.

Studies in general information systems implementation and diffusion areas have articulated and tested differences across stages of the innovation decision process. For example, innovation diffusion research postulates that many different outcomes are of interest in technology adoption [31, 57, 73], and IT implementation process can be decomposed into six stages: initiation, adoption, adaptation, acceptance, routinization, and infusion [31]. Most of TAM studies, however, mainly focused on cross-sectional investigation [58], while ignoring the temporal dimension of the adoption process [54]; consequently, these studies may not find causal linkage between research variables [38]. As observed by Rogers (1983), "the innovation decision process leading to institutionalization of usage may be conceptualized as a temporal sequence of steps through which an individual passes from initial knowledge of an innovation, to forming a favorable or unfavorable attitude towards it, to a decision to adopt or reject it, to put the innovation to use, and to finally seeking reinforcement of the adoption decision made". This problem is particularly pronounced when initial adoption decision is not well-formed; for example, perceived usefulness and ease of use based on vendor's demonstration of a product may deviate largely from perceptions based on a person's first hand experiences. True values of an IT, however, would only derive from its long-term, sustainable use as an integral part of end users' work process.

There are a number of reasons why the importance of temporal dynamics has been recognized, while research on this topic is still scarce. First, longitudinal studies tracing behavior changes are difficult to conduct, hence cross-sectional survey has become the predominant method. Cross-sectional surveys, however, only provide a snapshot of IT acceptance at a given time. They do not enable researchers to understand the time order of the variables, nor do they provide a complete picture of the time-dependent changes in behavior caused by IT use [74]. Even if surveys are administrated at multiple stages (pre- and post-implementation, or intermittently at month 1, 2, and 3, etc), it falls short in depicting a panorama of entire technology acceptance cycle, that is, from the formulation of initial adoption decision, to a stage when sustainable use is achieved. Second, TAM studies emphasize predicting a person's future usage based on his or her current beliefs, while "future" is highly ambiguous, which may result in mixed findings or equivocal interpretation of observations. As Davis et al (1989) indicated in an empirical study comparing TRA with TAM, both models revealed interesting developmental changes over time. For example, PU had increased influence on BI from time 1 to time 2. This fluctuation was interpreted as a result of converging beliefs from time 1 to time 2, that is, general usefulness constructs provided a somewhat better explanation of behavioral intension at time 1, while the consequent learning process converged the specific usefulness constructs to the general usefulness constructs, so that at time 2 PU yielded an increased influence. Effect of belief convergence may be a factor. However, solely attributing this fluctuation to belief convergence can be problematic because PU and BI measured at different stages may not represent the same psychological state.

In summary, all intention models, including TRA, TAM, and TRB, have an underlying supposition that behavioral intention predicts behavior. It is evident that intent to use an IT at point X predicts use behavior at point

$X + e$ within immediate temporal proximity, however there is no guarantee that this usage would be sustainable. Longitudinal studies that examine this evolving process, i.e., developmental patterns of IT acceptance behavior, is therefore highly desired.

4.3.5 Behavioral Intention, Actual Behavior, and Performance Gains

"I will do it" (behavioral intention) and "I have been doing it" (behavior) are certainly distinctive states. Albeit insufficient, behavioral intention is usually an antecedent of actual behavior [96] and in its own right generates interesting research questions. However "I want to but I cannot do it" is a common scenario in IT adoption, real or unreal. Factors such as lack of extrinsic incentives and intervening environmental conditions may impede a person's conducting an intended action, resulting in broken link between intention and behavior. Unfortunately, many studies focus solely on examining the causality between PU, PEoU, and behavioral intention (especially when actual use behavior is not measurable or no reliable measurement metrics exist), ignoring other factors that can be more fundamental determinants in formulating decisions of adopting or rejecting an IT. This problem has been attenuated in newer versions of TAM and other acceptance models, in which moderating and external factors are included to capture some of the potential contextual influences. For example, perceived control beliefs and facilitating conditions are present in TBP and UTAUT, respectively, postulated to have direct influence on behavior, and/or an indirect influence via behavioral intention [5,6,97]. Researchers also call for methodological pluralism: more relevant theories and methods, such as social network analysis, qualitative assessments, triangulation methods, diffusion of innovation theory, task-technology fit, and social cognitive theory may be used to further enhance understanding of technology acceptance [54,58,63].

Whether actual use of an IT would lead to anticipated outcomes (enhanced performance, improved productivity, etc) is a question that needs further investigation. TAM and its variants only address part of the chain in IT investment, acceptance, use, and returns. Many facets of the impact due to use of IT remains to be explored. As Davis et al. (1989) have suggested, "practitioners and researchers should not lose sight of the fact that usage is only a necessary but not sufficient, condition for realizing performance improvements due to information technology" (Davis et al. 1989, page 1000).

4.4 Assessing Physician IT Acceptance

The present study concerns physicians' acceptance behavior of a clinical decision support system. As Anderson et al. (1994) pointed out "despite the fact that they are technologically sound, more than half of medical information systems fail because of user and staff resistance" (page 1). The majority of CDSS evaluation studies have focused on recommendation quality, i.e., accuracy and relevancy of computer-generated advisories; relatively few have examined user acceptance issues in realistic, routine settings [53], and even fewer have applied the established acceptance theories from social psychology or information systems research [28].

4.4.1 Professional Context of Physicians

Whether TAM and other acceptance models are appropriate in studying physicians' IT acceptance behavior remains to be explored. There are a number of notable reasons why TAM or TAM-based approaches need to be used with caution in this context. First, physicians receive privileges such as autonomy, prestige, and institutional power [78]. Given the values associated with these privileges, physicians' decisions about accepting or rejecting an IT may be congruent to very distinctive considerations compared with other professionals or non-professionals with whom the prevalent models have proved validity. "Although information technologies in general may improve physicians' performances, certain types of information technologies, such as decision support systems and expert systems may, at the same time, undermine the monopoly of medical knowledge and reduce non-measurability of physicians' job performance. As a result, physicians' professional autonomy may be eroded and professional dominance compromised" [84]. Additional factors that are critical in formulating physicians' acceptance behavior, therefore, shall be examined and incorporated into the established frameworks.

Second, the well established constructs of TAM and other theoretical models may no longer carry the same conceptualizations that they represent in other settings. For example, PU refers to "the degree to which a person believes that using a particular system would enhance his or her job performance" [33, 35]. This definition is equivocal given the "job performance" of physicians is hard to measure and performance evaluations are less critical to their survival and success. PEoU, again, may be hard to define because, while workflow integration has been recognized as the most critical feature for health information systems' success (e.g. Kawamoto et al., 2005), an easy-to-learn system does not necessarily imply one that is easily integrated into physicians' traditional workflow and practice style. New constructs introduced in TAM's successive models, such as output quality, result demonstrability, performance expectancy, are all thrown into question in a similar manner.

Third, validity and reliability of the survey instruments inherited from TAM studies need to be vigorously tested in the professional context of physicians. As Chau and Hu (2001) commented, "instruments that have been developed and repeatedly tested in previous studies involving end users and business managers in ordinary business settings may not be equally valid in a professional setting, resulting in low reliability values of some scales". Some of the TAM's original PU measurement items, such as "using this technology in my job would enable me to accomplish tasks more quickly" and "using this technology would make it easier to do my job", do not seem to be congruent for many healthcare decision support applications that mainly aim to improve quality of care instead of increasing efficiency significantly. Telemedicine technologies, for example, are prone to be used as an alternative channel of care delivery, instead of helping "accomplish tasks more quickly". The psychometrical properties of all established instrument items therefore need to be carefully reevaluated. Ultimately, a new or revised set of instruments shall be developed to specifically address issues in physicians' acceptance of health information technologies.

4.4.2 Social and Organizational Issues

Healthcare organizations represent complex characteristics. Unlike many business applications for which individual users have full control over his or her own use decision and behavior, adoption and utilization of healthcare information technologies largely depends on the organizational contexts where the person is situated. For example, telemedicine technologies cannot be utilized if patients refuse or lack the capacity to cooperate; computerized physician order entry systems cannot be utilized until all parties (pharmacy, physician offices) agree to closely collaborate.

Adoption and acceptance of computer-based information systems usually involve three underlying structural aspects of an organization: *interdependency*, *interaction*, and *integration*. In healthcare, these organizational characteristics become particularly salient [10]. Fundamental tensions exist across occupational and departmental boundaries because of interdependency [92], resulting in conflict and struggles for power and access to resources [70]. Meanwhile, interaction within and across boundaries represents opportunities for gaining understanding of specialized but interdependent information and sharing of resources, leading to cooperation and coordination (e.g., Aydin, 1989). Such interaction has been demonstrated to determine medical practitioners' attitude towards and choice of innovations, for example new drugs (e.g., Coleman, 1966 and Burt, 1987) and computer applications (e.g., Anderson et al., 1987). Integration, as a consequence of introduction of information technologies, blurs the existing occupational and departmental boundaries, thus affecting the complex and subtle interplay of structural differences involving interdependencies and interaction [52]. For example Lundsgaarde et al. (1981) found that nurses and ancillary personnel readily accepted and used a computerized medical record system because it led to an expansion of their professional roles, whereas physicians refused to cooperate because it was more time-consuming than the manual system and they feared that it would disrupt traditional staff relations.

TAM and other prevalent acceptance models have a tacit focus on individual users' IT adoption and acceptance behavior. As a result they are inadequate for addressing relevant behaviors in more complex organizational settings. Factors associated with these aspects often outweigh an individual's attitudes and behavioral intention, leading to altered behavior. Theories and methods at organization level shall be employed to shape a better understanding of how a person forms and alters his or her IT acceptance behavior congruent to the job role, people around him or her, and the organizational context the person is situated.

4.4.3 Review of the Existing Studies

To date, only a few studies have used TAM, TRA, TRB, or similar theories to examine physicians' IT acceptance behavior. These studies are summarized in Table 4.1. As Table 4.1 reveals, these studies show mixed and inconclusive findings. Some argued the prevalent acceptance models may not be appropriate in the professional context of physicians, which necessitates additional constructs to be included. For example Succi and Walter (1999) introduced "PU towards professional status" and Pare et al. (2005) proposed "psychological ownership" in order to account for factors such as physicians' professionalism. Other studies showed that the existing models were partially adequate and applicable, nevertheless some findings were not in agreement with the previous studies conducted in traditional business settings. For example Hu et al. (1999) reported that PU and attribute towards telemedicine technology together accounted for 44% of the variance of use intention, however PEoU was found to have little influence. Chau and Hu (2001, 2002a, 2002b) conducted a model comparison study using the same dataset. They also found that PU and attitude were fundamental determinants of behavioral intention, while PEoU and SN did not seem to play a role. Another study conducted by Chrismar and Wiley-Patton (2002, 2003) tested TAM2 in the context of physicians' use intention of internet-based health applications. They concluded that although TAM2 overall explained 59% of intention variance, postulated effects of several key constructs, such as PEoU and SN, were not supported.

4.4.4 Limitations of the Existing Studies

Compared to the vast body of studies on user acceptance in other domains, research on physicians' IT acceptance is still quite preliminary. As shown in Table 4.1, the existing studies mainly employed one-time, cross-sectional study design; none used longitudinal investigation to trace evolvement of acceptance behavior. More problematically, the majority of these studies assessed physicians' attitudes and use intention for health information technologies (e.g. telemedicine technology or internet-based health applications) that were either not yet implemented or at early stage of development of the studied user population. Without a tangible system already deployed or within an anticipated timeframe of implementation, many key constructs such as PEoU, SN, output quality, and results demonstrability are not readily measurable. The

Table 4.1. Summaries of TAM Studies on Physicians' IT Acceptance

Study	Theories	Context of Study	Study Design	Main Findings
Pare et al. (2005)	TAM	91 out of 125 physician users of a physician order entry system responded to a mail survey.	Cross-sectional	A newly introduced construct, psychological ownership, was found to be positively associated with physicians' PU and PEoU of the CPoE system.
Chang et al. (2004)	TAM	23 nurses and 6 physicians were surveyed on their use intention of mass-gathering emergency medical service PDA support systems.	Cross-sectional	Satisfactory PU and PEoU were reported, which related to high use intention of the technology.
Chismar and Wiley-Patton (2002, 2003)	TAM2	89 out of 205 pediatricians responded to a mail questionnaire concerning their use intention of internet based health applications	Cross-sectional	TAM2 explains 59% of the variance in behavioral intention. Job relevance and output quality were found to have significant influence on PU, whereas PEoU and social processes (SN and image) were found to have no effect.
Croteau and Vieru (2002)	TAM, DIT	127 out of 390 questionnaires distributed via mail received valid response. Physician respondents were chosen from two groups: one from urban institution with little exposure of the technology, the other from rural areas who had some prior experience.	Cross-sectional	A revised TAM model is proposed and tested. Additional constructs included situation support, perceived confidence, image, and perceived voluntariness of use. Different, inconclusive results were reported across two groups. PU was consistently found to be the most influential factor, whereas other constructs showed little or no effect.
Chau and Hu, (2001, 2002a, 2002b)	TAM, TPB, and a revised DTFB	Based on the study conducted by Hu et al. (1999)	Cross-sectional	PU, attribute and PBC were found to have significant influence on behavioral intention, and compatibility had positively influences PU. PEoU and SN were reported no effect. DTPB did not substantially increase the predictive and explanatory power of TAM or TPB.

Continued from previous page

Dixon and Stewart (2000)	ITAM	101 out of 187 general and family physicians responded to a mail survey concerning adoption of health information technologies.	Cross-sectional	The authors proposed and tested a new model called Information Technology Adoption Model (ITAM), which is largely analogous to TAM. Physicians were stratified into three usage groups based on their self-assessments. A number of variables were tested using ANOVA and post hoc Tukey HSD test across these usage groups. Significant inter-group differences were found in respondents' intent to use, interest in using, PU and PEoU, finesse, knowledge, and non-clinical hours.
Hu et al. (1999)	TAM	408 out of 1728 mail questionnaire received valid response. Physician respondents were chosen from pre-selected specialties who practiced in a number of Hong Kong public tertiary hospitals. The survey assessed respondents' use intention of telemedicine technology.	Cross-sectional	PU and attitude together accounted for 44% of the variance of behavioral intention of adopting telemedicine technology. PEoU was found to have little influence.
Succi and Walter (1999)	TAM	Theoretical discussion	Not applicable	The authors proposed a new theoretical construct, perceived usefulness towards professional status, that captures professionals' attitudes about the impact of information technologies on their professional status.

results obtained from survey questions such as "I would use it when it becomes available", therefore, can deviate significantly from future perceptions when first hand experience of the technology becomes available. This common study design may explain why PEoU and SN were consistently found to be irrelevant. Types of information technologies studied are also limited: over half of the studies were conducted in the context of telemedicine acceptance, other prevalent (and critical) health informatics applications, such as EMR, CDSS, CPoE, have not yet been studied.

These studies also have a number of other notable problems. First, the population of physicians participating in mail surveys may suffer from self-selection bias. In turn, physicians who respond may be those who are more interested in the technologies in question. Second, while several new constructs have been introduced to address the unique professional characteristics and work environments of physicians, there is a lack of theoretical foundation for inclusion of new constructs, e.g., perceived confidence, finesse, etc. It is not clear why these factors may have an influence over and above the traditional set of well formulated constructs, and how these new variables relate to the established ones. Substantive knowledge and theoretical justification of inclusion of these additional constructs is required, that is, how to interpret their effects, how to operationalize these constructs, whether more potentially influential factors are left out, etc. Unfortunately these questions were not adequately addressed in the previous studies. Finally, behavioral intention remains the main dependent variable in these studies; none involved investigation in actual acceptance behavior. As mentioned earlier, while behavioral intention is an important antecedent, it is not the sole determinant of actual behavior. In examining physicians' IT acceptance this problem is more salient. For example, in a study that assessed the relationship between practitioners' behavioral intention to comply with clinical practice guidelines and their actual guideline compliance, only small correlations were found [65]. This problem is magnified when behavioral intention is formed from general perceptions of a targeted behavior, rather than deriving from a person's first hand experience, which is usually the case in the existing studies.

These studies described above consistently reported that PEoU and SN had no influence on PU or use intention of information technologies by physicians. The controversial role of PEoU has been a long-lasting issue in IT acceptance research[17]. Given that the technologies being studied were often not in use, thus the PEoU measures may not be accurate. Moreover, physicians are supported by other clinical personnel. Certain technologies (e.g., telemedicine) may not be directly used by physicians. Conclusions drawn based on surveying physicians alone can be misleading.

---

[17] Influence of PEoU largely depends on inherent nature of the information systems under investigation as well as respondents' role in use of the systems. See Lee et al. (2003), page 759–760.

The lack of statistical significance of SN is striking and contradicts the findings from a large volume of medical research. As mentioned earlier, the structure of communication networks among medical practitioners significantly affect the rate of adoption and diffusion of medical technology [10]. There is also long-standing evidence that personal contacts among physicians play an important role in the diffusion of new treatments, and choices are subject to a variety of implicit or explicit social pressures. For example studies from medical research have revealed that subjective norm (e.g., colleagues' approval or disapproval) strongly influences physicians' behavior, such as new drug adoption (e.g., Coleman, 1966, Burt, 1987, and Gaither et al., 1996), prescription behavior (e.g., Legare, 2005), and clinical practice and guideline coherence (e.g., Stross and Bole, 1980 and Eisenberg et al., 1983). In the context of telemedicine that were commonly examined in TAM studies, Gagnon et al. (2003) showed that using an adapted interpersonal behavior theory, a composite normative factor (comprising personal as well as social norms) was found to be the most significant psychosocial determinant of physicians' intention to adopt telemedicine technologies. These evidence collectively shows that the "no effect" result of subject norm derived from TAM-based models is equivocal. It is probably because this theoretical construct was not applicable or improperly operationalized in these studies.

4.5 Conclusions

Adequate user acceptance is a critical prerequisite for any technology success. Even if a healthcare IT application is optimized as much as technically possible, practitioners may not acknowledge that use of the system would add value to their medical practice and thus may be reluctant to incorporate it into their daily routine. This chapter highlights the importance of user acceptance in the health IT diffusion process and reviews major evaluation methodologies—notably the technology acceptance model (TAM) which draws on well established social psychology theories and has become the prevalent method of studying user acceptance in the area of information system research.

While TAM has been extensively tested and proven to be a powerful theory, its limitations have also been recognized. This chapter analyzes these limitations and discusses the potential pitfalls of applying TAM in empirical setting, namely model operationalization, self-reported measures, type of information systems studies, temporal dynamics, and behavioral intention, actual behavior, and performance gains. This chapter also examines the contextual issues of applying TAM to the technology adoption issues in healthcare, with a review of available studies. In summary, TAM and TAM-based models remain to be useful tools in studying the user acceptance of healthcare IT applications, however, special cautions—particularly those concerning the applicability of the existing models in this new domain—should be used.

References

1. D. ADAMS, R. NELSON, AND P. TODD, *Perceived usefulness, ease of use, and usage of information technology: a replication*, MIS Quarterly, 16 (1992), pp. 227–247

2. R. AGARWAL AND E. KARAHANNA, *Time flies when you're having fun: cognitive absorption and beliefs about information technology usage*, MIS Quarterly, 24 (2000), pp. 665–694

3. R. AGARWAL AND J. PRASAD, *The role of innovation characteristics and perceived voluntariness in the acceptation of information technologies*, Decision Sci, 28 (1997), pp. 557–582

4. ———, *Are individual differences germane to the acceptance of new information technologies?*, Decision Sci, 30 (1999), pp. 361–391

5. I. AJZEN, *From intentions to actions: a theory of planned behavior*, in Springer series in social psychology, J. Kuhl and J. Beckmann, eds., Springer, Berlin, 1985, pp. 11–39

6. I. AJZEN, *The theory of planned behavior*, Organizational Behavior and Human Decision Processes, 50 (1991), pp. 179–211

7. I. AJZEN AND M. FISHBEIN, *Understanding Attitudes and Predicting Social Behavior*, Prentice-Hall, Englewood Cliffs, NJ, 1980

8. D. ALBARRACIN, B. T. JOHNSON, M. FISHBEIN, AND P. A. MUELLERLEILE, *Theories of reasoned action and planned behavior as models of condom use: a meta-analysis*, Psychol Bull, 127 (2001), pp. 142–161

9. J. ANDERSON, S. JAY, H. SCHWEER, M. ANDERSON, AND D. KASSING, *Physician communication networks and the adoption and utilization of computer applications in medicine*, in Use and Impact of Computers in Clinical Medicine, J. G. Anderson and S. J. Jay, eds., Springer-Verlag, New York, 1987, pp. 185–199

10. J. G. ANDERSON, C. E. AYDIN, AND S. J. JAY, *Evaluating Health Care Information Systems*, SAGE Publications, Thousand Oaks, CA, 1994

11. C. J. ARMITAGE AND M. CONNER, *Efficacy of the theory of planned behaviour: a meta-analytic review*, Br J Soc Psychol, 40 (2001), pp. 471–499

12. C. E. AYDIN, *Occupational adaptation to computerized medical information systems*, J Health Soc Behav, 30 (1989), pp. 163–179

13. A. BANDURA, *Social Foundations of Thought and Action: A Social Cognitive Theory*, Prentice Hall, Englewood Cliffs, New Jersey, 1986

14. ———, *Social cognitive theory*, in Annals of Child Development Vol. 6, R. Vasta, ed., Jai Press Ltd, Greenwich, CT, 1989, pp. 1–60

15. H. BARKI AND J. HARTWICK, *Measuring user participation, use involvement, and user attitude*, MIS Quarterly, 18 (1994), pp. 59–82

16. L. BECK AND I. AJZEN, *Predicting dishonest actions using the theory of planned behavior*, J of Research and Personality, 25 (1991), pp. 285–301

17. E. BLAIR AND S. BURTON, *Cognitive process used by survey respondents to answer to behavioral frequency questions*, J of Consumer Research, 14 (1987), pp. 280–288

18. N. M. BRADBURN, J. HUTTENLOCHER, AND L. HEDGES, *Telescoping and temporal memory*, in Autobiographical memory and the validity of retrospective reports, N. Schwarz and S. Sudman, eds., Springer Verlag, New York, 1994, pp. 203–216

19. N. M. BRADBURN, L. J. RIPS, AND S. K. SHEVELL, *Answering autobiographical questions: the impact of memory and inference on surveys*, Science, 236 (1987), pp. 157–161
20. R. S. BURT, *Social contagion and innovation: Cohesion versus structural equivalence*, The American Journal of Sociology, 92 (1987), pp. 1287–1335
21. P. CHANG, Y. S. HSU, Y. M. TZENG, Y. Y. SANG, I. C. HOU, AND W. F. KAO, *The development of intelligent, triage-based, mass-gathering emergency medical service PDA support systems*, J Nurs Res, 12 (2004), pp. 227–236
22. P. Y. K. CHAU AND P. J. H. HU, *Information technology acceptance by professionals: A model comparison approach*, Decision Sciences, 32 (2001), pp. 699–719
23. ———, *Examining a model of information technology acceptance by individual professionals: An exploratory study*, J of Management Information Systems, 18 (2002), pp. 191–229
24. ———, *Investigating healthcare professionals' decisions on telemedicine technology acceptance: An empirical test of competing theories*, Information and Management, 39 (2002), pp. 297–311
25. C. CHEN, M. CZERWINSKI, AND R. MACREDIE, *Individual differences in virtual environments - introduction and overview*, J of the American Society for Information Science, 51 (2000), pp. 499–507
26. W. W. CHIN, *The measurement and meaning of IT usage: reconciling recent discrepancies between self reported and computer recorded usage*, in Proceedings of the Administrative Sciences Association of Canada, Information Systems Division, Montreal, Quebec, Canada, 1996, pp. 65–74
27. W. G. CHISMAR AND S. WILEY-PATTON, *Test of the technology acceptance model for the internet in pediatrics*, Proc AMIA Symp, (2002), pp. 155–159
28. ———, *Does the extended technology acceptance model apply to physicians*, in HICSS '03: Proceedings of the 36th Annual Hawaii International Conference on System Sciences (HICSS-36), Washington, DC, USA, 2003, IEEE Computer Society, p. 160
29. K. E. CLARKE AND A. AISH, *An exploration of health beliefs and attitudes of smokers with vascular disease who participate in or decline a smoking cessation program*, J Vasc Nurs, 20 (2002), pp. 96–105
30. J. COLEMAN, E. KATZ, AND H. MENZEL, *Medical Innovation: A Diffusion Study*, Bobbs-Merrill, New York: NY, 1966. 2nd
31. R. B. COOPER AND R. W. ZMUD, *Information technology implementation research: a technological diffusion approach*, Manage. Sci., 36 (1990), pp. 123–139
32. F. D. DAVIS, *A technology acceptance model for empirically testing new end-user information systems: theory and results*, PhD thesis, Sloan School of Management, Massachusetts Institute of Technology, 1986
33. ———, *Perceived usefulness, perceived ease of use, and user acceptance of information technology*, MIS Quarterly, 13 (1989), pp. 319–342
34. F. D. DAVIS, *User acceptance of information technology: system characteristics, user perceptions and behavioral impacts*, Int J Human-Computer Studies, 38 (1993), pp. 475–487
35. F. D. DAVIS, R. P. BAGOZZI, AND P. R. WARSHAW, *User acceptance of computer technology: a comparison of two theoretical models*, Management Sci, 35 (1989), pp. 982–1003

36. T. J. DeMaio, *Social desirability and survey measurement: a review*, in Surveying subjective phenomena, C. F. Turner and E. Martin, eds., Russell Sage, New York, 1984, pp. 257–281

37. D. R. Dixon and M. Stewart, *Exploring information technology adoption by family physicians: survey instrument valuation*, Proc AMIA Symp, (2000), pp. 185–189

38. W. J. Doll and M. U. Ahmed, *Managing user expectations*, J of Systems Management, 34 (1983), pp. 6–11

39. J. M. Eisenberg, D. S. Kitz, and R. A. Webber, *Development of attitudes about sharing decision-making: a comparison of medical and surgical residents*, J Health Soc Behav, 24 (1983), pp. 85–90

40. M. Fishbein and I. Ajzen, *Beliefs, Attitude, Intention and Behavior: An Introduction to Theory and Research*, Addison-Wesley, Reading, MA, 1975

41. M. P. Gagnon, G. Godin, C. Gagne, J. P. Fortin, L. Lamothe, D. Reinharz, and A. Cloutier, *An adaptation of the theory of interpersonal behaviour to the study of telemedicine adoption by physicians*, Int J Med Inform, 71 (2003), pp. 103–115

42. C. A. Gaither, R. P. Bagozzi, F. J. Ascione, and D. M. Kirking, *A reasoned action approach to physicians' utilization of drug information sources*, Pharm Res, 13 (1996), pp. 1291–1298

43. C. L. Gatch and K. D, *Predicting exercise intentions: the theory of planned behavior*, Research Quarterly For Exercise and Sport, 61 (1990), pp. 100–102

44. D. Gefen and D. Straub, *Gender difference in the perception and use of E-Mail: an extension to the technology acceptance model*, MIS Quarterly, 21 (1997), pp. 389–400

45. G. Godin, P. Valois, L. Lepage, and R. Desharnais, *Predictors of smoking behavior: an application of Ajzen's theory of planned behavior*, British Journal of Addition, 87 (1992), pp. 1335–1343

46. M. S. Hagger, N. L. D. Chatzisarantis, and S. J. H. Biddle, *Meta-analysis of the theories of reasoned action and planned behavior in physical activity: an examination of predictive validity and the contri-bution of additional variables*, J of Sport and Exercise Psychol, 24 (2002), pp. 3–32

47. C. Hartley, M. Brecht, P. Pagerly, G. Weeks, A. Chapanis, and D. Hoecker, *Subjective time estimates of work tasks by office workers*, J of Occupational Psychology, 50 (1977), pp. 23–36

48. P. Hu, P. Y. K. Chau, O. R. L. Sheng, and K. Y. Tam, *Examining the technology acceptance model using physician acceptance of telemedicine technology*, J of Management Information Systems, 16 (1999), pp. 91–112

49. G. S. Hubona and P. H. Cheney, *System effectiveness of knowledge-based technology: the relationship of user performance and attitudinal measures*, in HICSS '94: Proceedings of the 27th Annual Hawaii International Conference on System Sciences (HICSS-27) Volume 4, Washington, DC, USA, 1994, IEEE Computer Society, pp. 532–541

50. M. Igbaria, T. Guimaraes, and G. B. Davis, *Testing the determinants of microcomputer usage via a structural equation model*, J of MIS, 11 (1995), pp. 87–114

51. M. Igbaria, J. Iivari, and H. Maragahh, *Why do individuals use computer technology? a Finnish case study*, Information & Management, 29 (1995), pp. 227–238

52. B. KAPLAN, *Addressing organizational issues into the evaluation of medical systems*, Academy of Management Journal, 4 (1997), pp. 94–101

53. ———, *Evaluating informatics applications - some alternative approaches: theory, social interactionism, and call for methodological pluralism*, Int J Med Inform, 64 (2001), pp. 39–56

54. E. KARAHANNA, D. W. STRAUB, AND N. L. CHERVANY, *Information technology adoption across time: a cross-sectional comparison of pre-adoption and post-adoption beliefs*, MIS Quarterly, 23 (1999), pp. 183–213

55. Y. KASHIMA, C. GALLOIS, AND M. McCAMISH, *Theory of reasoned action and cooperative behavior: it takes two to use a condom*, British Journal of Social Psychology, 32 (1993), pp. 227–239

56. H. C. KELMAN, *Compliance, identification, and internalization: three processes of attitude change?*, J of Conflict Resolution, 2 (1958), pp. 51–60

57. T. H. KWON AND R. W. ZMUD, *Unifying the fragmented models of information systems implementation*, in Critical issues in information systems research, R. J. Boland and R. A. Hirschheim, eds., John Wiley & Sons, Inc., Chichester, England, 1987, pp. 227–251

58. Y. LEE, K. A. KOZAR, AND K. R. T. LARSEN, *The technology acceptance model: past, present, and future*, Communications of the AIS, 12 (2003), pp. 752–780

59. F. LEGARE, G. GODIN, V. RINGA, S. DODIN, L. TURCOT, AND J. NORTON, *Variation in the psychosocial determinants of the intention to prescribe hormone therapy prior to the release of the Women's Health Initiative trial: a survey of general practitioners and gynaecologists in France and Quebec*, BMC Med Inform Decis Mak, 5 (2005), p. 31

60. H. P. LUNDSGAARDE, P. A. FISCHER, AND D. J. STEELE, *Human problems in computerized medicine*, in Publications in Anthropology, 13, The University of Kansas, Lawrence, KS, 1981

61. Y. MALHOTRA AND D. F. GALLETTA, *Extending the technology acceptance model to account for social influence: theoritical bases and empirical validation*, in HICSS '99: Proceedings of the 32nd Annual Hawaii International Conference on System Sciences (32), vol. 1, Washington, DC, USA, 1999, IEEE Computer Society, p. 1006

62. T. A. MASSARO, *Introducing physician order entry at a major academic medical center: I. impact on organizational couture and behavior*, Acad Med, 68 (1993), pp. 20–25

63. K. MATHIESON, *Predicting user intentions: comparing the technology acceptance model with the theory of planned behavior*, Information Systems Research, 2 (1991), pp. 173–191

64. K. MATHIESON, E. PEACOCK, AND W. W. CHIN, *Extending the technology acceptance model: the influence of perceived user resources*, ACM SIGMIS Database, 32 (2001), pp. 86–112

65. S. K. MAUE, R. SEGAL, C. L. KIMBERLIN, AND E. E. LIPOWSKI, *Predicting physician guideline compliance: an assessment of motivators and perceived barriers*, The American J of Managed Care, 10 (2004), pp. 382–391

66. E. MAYO, *The Human Problems of an Industrial Civilization*, MacMillan, New York, NY, USA, 1933

67. G. C. MOORE AND I. BENBASAT, *Development of an instrument to measure the perceptions of adopting an information technology innovation*, Information Systems Research, 2 (1991), pp. 173–191

68. D. M. MORRISON, M. S. SPENCER, AND M. R. GILLMORE, *Beliefs about substance use among pregnant and parenting adolescents*, J Res Adolesc, 8 (1998), pp. 69–95

69. R. S. NICKERSON, *Why interactive computer systems are sometimes not used by people who might benefit from them*, Int J Human-Computer Studies, 51 (1999), pp. 307–321

70. M. H. OLSON AND H. C. LUCAS, *The impact of office automation on the organization: some implications for research and practice*, Commun ACM, 25 (1982), pp. 838–847

71. G. PARÉ, C. SICOTTE, AND H. JACQUES, *The effects of creating psychological ownership on physicians' acceptance of clinical information systems*, J Am Med Inform Assoc, 13 (2006), pp. 197–205

72. D. PARKER, A. S. R. MANSTEAD, S. G. STRADLING, AND J. REASON, *Intention to commit driving violations: An application of the theory of planned behavior*, J of Applied Psychology, 77 (1992), pp. 94–101

73. M. B. PRESCOTT AND S. A. CONGER, *Information technology innovations: a classification by it locus of impact and research approach*, SIGMIS Database, 26 (1995), pp. 20–41

74. E. M. ROGERS, *Diffusion of Innovations*, The Free Press, New York, 1983. 3rd ed

75. ———, *Diffusion of Innovations*, The Free Press, New York, 1995. 4th ed

76. V. SAMBAMURTHY AND C. W.W., *The effects of group attitudes toward alternative gdss designs on the decision-making performance of computer-supported groups*, Decision Sciences, 25 (1994), pp. 215–241

77. N. SCHWARZ AND D. OYSERMAN, *Asking questions about behavior: cognition, communication, and questionnaire construction*, American J of Evaluation, 22 (2001), pp. 127–160

78. A. SHARMA, *Professionals as agent: knowledge asymmetry in agengy exchanges*, Academy of Management Review, 22 (1997), pp. 758–798

79. B. H. SHEPPARD, J. HARTWICK, AND P. R. WARSHAW, *The theory of reasoned action: a meta-analysis of past research with recommendations for modifications and future research*, J of Consumer Behavior, 15 (1988), pp. 325–343

80. D. E. SICHEL, *The Computer Revolution: An Economic Perspective*, Brookings, Washington, DC, 1997

81. D. STRAUB, M. KEIL, AND W. BRENNER, *Testing the technology acceptance model across cultures: a three country study*, Information & Management, 33 (1997), pp. 1–11

82. J. STROSS AND G. BOLE, *Evaluation of a continuing education program in rheumatoid arthritis, arthritis and rheumatism*, Arthritis Rheum, 23 (1980), pp. 846–849

83. G. SUBRAMANIAN, *A replication of perceived usefulness and perceived ease of use measurement*, Decision Sciences, 25 (1994), pp. 863–874

84. M. J. SUCCI AND Z. D. WALTER, *Theory of user acceptance of information technologies: an examination of health care professionals*, in HICSS '99: Proceedings of the 32nd Annual Hawaii International Conference on System Sciences (HICSS-32), Los Alamitos, CA, 1999, IEEE Computer Society

85. H. SUN AND P. ZHANG, *The role of moderating factors in user technology acceptance*, Int J Human-Computer Studies, 64 (2006), pp. 53–78

86. E. B. SWANSON, *Information System Implementation: Bridging the Gap between Design and Utilization*, Irwin, Homewood, IL, 1988

87. B. Szajna, *Empirical evaluation of the revised technology acceptance model*, Management Sci, 42 (1996), pp. 85–92

88. S. Taylor and P. Todd, *Understanding information technology usage: a test of competing model*, Information Systems Research, 6 (1995), pp. 145–176

89. R. L. Thompson, C. A. Higgins, and J. M. Howell, *Personal computing: toward a conceptual model of utilization*, MIS Quarterly, 15 (1991), pp. 124–143

90. ———, *Influence of experience on personal computer utilization: testing a conceptual model*, Journal of Management Information Systems, 11 (1994), pp. 167–187

91. L. Tornatzky and K. Klein, *Innovation characteristics and innovation adoption implementation: a meta-analysis of findings*, IEEE Transactions on Engineering Management, 29 (1982), pp. 28–45

92. M. L. Tushman and T. J. Scanlan, *Boundary spanning individuals: their role in information-transfer and their antecedents*, Academy of Management Journal, 24 (1981), pp. 289–305

93. V. Venkatesh, *Determinants of perceived ease of use integrating control, intrinsic motivation, and emotion into the technology acceptance model*, Information Systems Research, 11 (2000), pp. 342–365

94. V. Venkatesh and F. D. Davis, *A model of the antecedents of perceived ease of use development and test*, Decision Sciences, 27 (1996), pp. 451–481

95. ———, *A theoretical extension of the technology acceptance model: four longitudinal field studies*, Management Sci, 46 (2000), pp. 186–204

96. V. Venkatesh and M. G. Morris, *Why don't men ever stop to ask for directions? gender, social influence, and their role in technology acceptance and usage behavior*, MIS Quarterly, 24 (2000), pp. 115–139

97. V. Venkatesh, M. G. Morris, G. B. Davis, , and F. D. Davis, *User acceptance of information technology: toward a unified view*, MIS Quarterly, 27 (2003), pp. 425–478

5

Current Perspectives on PACS
and a Cardiology Case Study

Carlos Costa, Augusto Silva, and José Luís Oliveira

Universidade de Aveiro, DETI/IEETA, Portugal
{ccosta, asilva, jlo}@det.ua.pt

Abstract. Since the first experiments on digital medical imaging, Pictures Archiving and Communication Systems (PACS) have been gaining acceptance along healthcare practitioners. PACS based infrastructures are currently being driven by powerful medical applications that rely completely on the seamless access to images' databases and related metadata. New and demanding applications such as study co-registration and content based retrieval are already driving PACS into new prominent roles.
In this chapter we will revise the major key factors that have promoted this technology. We will then present our own solution for a Web-based PACS and the results achieved by its use on a Cardiology Department. We will finally consider future applications that are pushing developmental research in this field.

5.1 Introduction

Over the last decade, the use of digital medical imaging systems has increased greatly in healthcare institutions. Today it is one of the most valuable tools supporting medical profession in both decision making and treatment procedures.

The research and industry efforts to develop from the beginning both centralized and standalone medical imaging machines to a evolve grid of networked imaging resources have been major driving forces towards the acceptance of the Picture Archiving and Communication System (PACS) concept.

Practitioner's satisfaction has arisen as faster and more generalized access to image data was enabled by PACS implementations. It reduced the costs associated with the storage and management of image data and it also increased both the intra and inter-institutional portability of data. One of the most important benefits of the digital medical image is that it allows the widespread sharing and remote access to medical data by outside institutions.

C. Costa et al.: *Current Perspectives on PACS and a Cardiology Case Study*, Studies in Computational Intelligence (SCI) **65**, 79–108 (2007)
www.springerlink.com

PACS presents an opportunity to improve cooperative workgroups taking place either within or with other healthcare institutions. The most important contribution to the exchange of structured medical imaging data was the establishment of the DICOM Standard (Digital Imaging and Communications in Medicine), in 1992. Currently, almost all medical imaging equipment manufacturers provide embedded DICOM (version 3) digital output in their products.

The dissemination of medical images in electronic format allows the appearance of a large variety of post-processing techniques that extract more and better information from the data. The clinical practice is full of examples of image processing applications that rely on PACS resources. For instance, Computer-Aided Diagnosis (CAD), Image-Assisted Surgery Systems (IASS) and multimodality studies are some of the applications that retrieve images from PACS. It may be done intensively, and there is the possibility to benefit from current medical research by using modern indexing and retrieval strategies.

Medical image data can be generated in practically any healthcare institution, even by those with limited human or financial resources. However, modeling and quantitative analysis tools are especially expensive with respect to software costs and computational power requirements. Specialized physicians are usually concentrated in a limited number of medical centers. This suggests that new approaches should be developed and could include the establishment of a trans-institutional grid of technical and human resources. The grid could include computational facilities such as data processing, analysis and reporting. In this area, PACS are playing an important role in the intra and inter institutional clinical environment. However, many issues remain unsolved and they are mainly related to data volumes and to the need to provide interoperability between heterogeneous systems. Another requirement is to provide wide access to all wishing to use the system. For instance, a problematic barrier is the handling of dynamic image modalities (films) such as cardiac ultrasound (US) and X-Ray angiography (XA). The date scanning rate and the data volume associated with these modalities pose several problems to the design and deployment of PACS.

In this chapter we will present the work we have done in a cardiac imaging environment during the last 5 years using a platform entitled HIMAGE PACS. The aim has been to promote medical image availability and greater accessibility to users. This system incorporates a new DICOM encoding syntax for data compression, that allows to have available on-line and virtually forever, all the examinations done in a Cardiology department. It has been used a diagnostic framework within and between institutions. In the final sections some advanced applications which have been enabled by PACS technologies will be presented and their role in a near future will be discussed.

5.2 Digital Medical Imaging

5.2.1 PACS

The term PACS encompasses technologies used for the acquisition, archive, distribution and visualization of a set of digital images using a computer network for diagnosis and revision in dedicated stations. Retrieving images from a PACS presents significant advantages over the traditional analogical systems based in film. To help in this task, the DICOM file header provides metadata information such as, for instance, modality, date, physician, accession number. These are often used on a daily basis for retrieving images. Needless to say that retrieval efficiency is several orders of magnitude better than the old traditional analog film practice. This allows an overall boost to productivity. In fact, the improved technology is no longer the only motive for the installation of PACS but also economical and management reasons start to make its way.

The implementation of the PACS clearly contributes to the increase in the productivity of health professionals. It increases both the efficiency and the quality of services provided to the patient. The physicians are now able to use tools that allow to remotely access patients' information, but also allows telemedicine, telework and collaborative work environments.

Digital image manipulation often relies on image processing techniques which are adapted for each of the image modality in order to increase the visibility of the clinical information and to obtain quantitative estimates of parameters. These parameters may act as helpful descriptors of a pathology. The amount of complementary information that can be unveiled with suitable modeling and quantification tools is surprisingly large. Consider for instance, how the subtle wrist fractures in Computed Radiography (CR) could be made more evident by the use of simple brightness and contrast manipulation or in the case of X-Ray angiography how the contrast dye filled vessels look more crisp with the use of linear enhancement filtering.

One can say that a long term problem that was often encountered in PACS was the difficulty in transferring image data between the equipment of different providers, in an easy way. Even inside the same Radiology Department, heterogeneity is the default situation between hardware and software vendors. If we have a clinical environment with imaging solutions provided from multiple manufacturers, all participants must establish a common communication protocol. This avoids the necessity to provide custom or proprietary interfaces. Currently, this interoperability service is universally provided by the Digital Imaging and Communications in Medicine (DICOM) standard).

5.2.2 DICOM

The ACR-NEMA (American College of Radiology - National Electrical Manufacturers Association) has been promoting the development of a set of recommendations and guidelines to allow the exchange of medical digital images between different equipment manufacturers. The result of this work is the DICOM standard (http://medical.nema.org/) which have had as direct consequence the development and expansion of PACS. Currently, the DICOM version 3.0 is divided into 18 parts and it includes 127 supplements. Some of these have been discontinued. Mainly it specifies items such as data formats, storage organization and communication protocols.

At the communications level, DICOM defines what and how information will be exchanged. The protocol proposes two steps. First, a negotiation phase must exist where many relevant aspects are agreed upon. For instance, the image modality that will be transmitted (XA, US, CT, etc.) and how this data will be encoded (little/big endian, un/compressed, etc.). Secondly, if the negotiation process is successfully completed, then effective data transfer between the two hosts takes place.

The images data structure and encoding are given in Part 5 of the norm [1]. A DICOM "Dataset" is formed by a variable number of sequential "Data Elements" (Fig. 5.1). Some of the elements are mandatory while other are optional.

The Elements, organized into groups and sub-groups, are composed by a predefined set of fields:

- Tag: A four bytes field used to identify the element. The first pair of bytes identifies the group number of the object and the last pair of bytes the number of the element. For example, the tag with the hexadecimal value (0010,0010) identifies the data element that contains the patient name.
- VR (Value Representation): This two bytes field is optional and its presence depends on the type of negotiated transfer syntax. The possible values in this field are defined in the dictionary of data. This is specified in Part 6 of the norm [2].

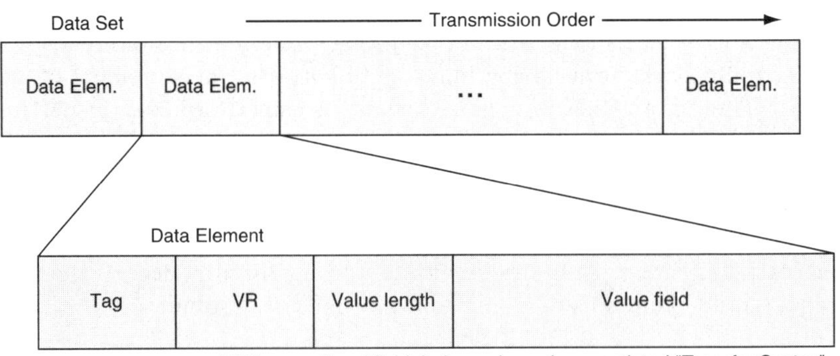

Fig. 5.1. DICOM "Data Set" e "Data Element"

Table 5.1. Example of DICOM Data Elements

Tag	VR	Length	Field	Comment
(0002,0010)	UI	18	1.2.840.10008.1.2	TransferSyntaxUID
(0008,0020)	DA	8	20010426	StudyDate
(0008,0060)	CS	2	US	Modality
(0010,0010)	PN	18	TEST PSEUDO SHEILA	PatientName
(0010,0020)	LO	10	ETT0000001	PatientID
(0028,0002)	US	2	3	SamplesPerPixel
(0028,0004)	CS	4	RGB	PhotometricInterpretation
(0028,0008)	IS	2	8	NumberOfFrames
(0028,0010)	US	2	512	Rows
(0028,0011)	US	2	480	Columns
(7fe0,0010)	OW	5898240	0000\0000\0000\... ...	PixelData

- Value Length: Contains the length of the field "value field", as the number of bytes. The length of this field can be of 16 or 32 bits, depending on the negotiated type of VR.
- Value Field: It contains the data element. This can be a simple element or another sub-group or set of sub-groups.

In Table 5.1 we find a list of some elements that compose a typical US DICOM file with an encoding syntax of the type "Default Transfer Syntax" (tag: 0002,0010). A Unique Identifier (UID) is used with value of 1.2.840.10008.1.2 as it is defined by the norm.

During the last decade we have assisted to a wide distribution of medical equipment with DICOM interfaces, conforming to a specified set of services. A large amount of diagnosis information supported by the PACS image databases became potentially available in many widely geographically distributed devices.

5.3 The Medical Imaging Laboratory

The traditional concept of a medical imaging environment has been directly associated with a Radiology Department. It is now more appropriate and flexible to think of image producing clusters as Medical Imaging Laboratories which can accommodate an ever growing number of modalities and related imaging services. From the DICOM point of view the idea of PACS is not restricted to the traditional Radiology environments. Now due to intensive image driven diagnostics and therapeutic procedures Cardiology was always a major candidate for PACS deployment.

5.3.1 Data Volumes

The volume of information generated in a specific imaging laboratory depends mainly on the image modalities and on the number of patients assisted. Each modality defines the type of the digital output while the image quality depends on adequate sampling in the spatial and temporal domains and also on the appropriate bit depth. Sampling and signal quantization is mostly driven by spatial and temporal resolutions and by signal to noise ratios. These factors are not independent as they are ultimately consequence of clinical and physical limitations. Those limitations include factors such as the dose in ionizing modalities or the magnetic field strength in Magnetic Ressonance Imaging (MRI). Digital data produced by each modality are thus typically available as spatial stacks of 2D arrays or temporal sequences of 2D arrays. The dimensions of each array, the pixel representation, the stack extent or the frame rate are the primary indicators for the estimation of data volumes.

Cardiac imaging is also often characterized by human factors. In fact, either in interventional coronary imaging thought cineangiography or in echocardiography the human operator has a role. In these cases, the operator must be educated and trained to do optimized captures in terms of the captured anatomic fields, and to restrict the period of time for the capture to that strictly necessary.

The digitization of the major dynamic cardiac modalities brought the concept of the digital cine-loop into daily practice. This means that, assuming that most of the sought for information is visible through one or a few more cardiac cycles, the operator may restrict the image acquisition period to a few seconds spanning the necessary cycles. This type of practice is often called "clinical image compression" and it certainly contributes to feasible storage strategies.

These multiple factors present obstacles to accurately estimate the data storage volumes required in cardiac imaging laboratories. Some average figures are illustrative:

- A modern 64-slice MultiDetector-row Computed Tomography (MDCT) examination can produce 3500 images which can require 1.7 GB of storage per study using typical matrix and pixel dimensions.
- A typical XA procedure can have 20 cine-loops with 8bits/pixel images on a 512 by 512 matrix and 90 frames. It represents a data volume of 450 MB.
- A typical echocardiography study with 10 runs and a few static images can need hundreds of MB.

In our work with echocardiography, we have experiences with equipments that, for similar clinical procedures, produce volumes of information with differences up to 3 to 1. This was mainly due to equipment specifications concerning matrix dimensions and ECG controlled acquisition. Even optimizing all of these factors, we are always dealing with tremendous volumes of information.

In most cases, the storage and the permanent availability in a departmental server of the entire image data produced are extremely difficult tasks. Considering the storage capacity and performance of current information technologies it is not easy to maintain on-line more that some months of information. This is even using the compression tools supported by the DICOM standard. That is if we intend to make use of compression factors that keep the quality of diagnosis for each modality at acceptable standards.

Medical imaging implies distribution and remote consultation on a timely effective basis which is supported by high communication costs over broadband platforms. The distribution or the remote access to the information in wide-scale network environments, such as the Internet, implies communication channels with reduced bandwidth and without assured QoS (Quality of Service). Image compression is thus a unique option if one intends to avoid hours of transmission for a single dynamic imaging procedure.

5.3.2 Image Compression

Although in the previous sections we have identified clinical techniques to reduce the size of captured image data, they, do not solve the "volume problem". In this context, digital image compression appears to be an indispensable tool for PACS.

Image compression methods can be broadly separated into two classes: lossless or reversible without losses, and lossy or reversible with losses. In the former class no information is lost and the original image can be reconstructed, pixel by pixel, from the compressed image. The latter method has some acceptable loss of information, when the image is decoded.

Reversible compression methods can only achieve modest reductions in storage requirement; typically, between 2:1 and 4:1 [3, 4]. Comparatively, irreversible compression can achieve substantially higher compression ratios without perceptible image degradation: ratios ranging from 10:1 to 50:1 or more [5], depending on the compression algorithm used and the required minimal quality of the compressed image. In the latter, the size of the compressed image is inversely influenced by the compression ratio. If the compression ratio increases, less storage space is required and faster transmission speeds are possible, but at the expense of image quality degradation [6].

It is known that irreversible compression does not allow the recover of the original image. The compression factor can be adapted to keep the intended image at clinically acceptable diagnosis quality given a particular modality and diagnostic task [7]. Several research studies have been done over the years to examine the impact of various compression ratios on image quality and observer performance for several diagnostic tasks. These have included lesion detection, identification and quantitative parameter estimation. The general trend shown by the results of these studies is that consensual compression ratios for a modality are rarely achieved as the observer performance is tightly coupled to each particular diagnostic task. There are guidelines often published by international scientific and professional organizations that recommend suitable algorithms and compression ratios for a broad range of diagnostic tasks. Using them, it should be possible to accomplish the task with a particular imaging modality (see [8–10] to XA and [11–13] to US).

Generally, for low compression ratios (8:1 or less), the quality loss is limited in such way that no visual impairment is perceptible. For diagnostic purposes the compression algorithm may in fact be classified as visually lossless [5]. To avoid legal consequences of an incorrect diagnosis based on an irreversible compressed image, many of these algorithms are not used by radiologists in primary diagnosis [Huang, 2002]. This is especially applicable in countries such as the USA and Canada.

Until recently, the only supported method of lossy compression within DICOM was JPEG [14]. This continues to be the mostly used method for many distinct imaging modalities. In January 2002 the ACR-NEMA approved the supplement 61 of DICOM standard [15] that contemplates the "JPEG2000 Transfer Syntaxes". This algorithm relies on robust wavelet methods that provide improved image quality especially at high compression ratios as well as multi-resolution storage and communication [16]. However, the JPEG algorithms have a handicap when applied to dynamic image modalities (i.e. cine-loops) as they only explore the intra-frame (spatial) redundancy. In 2004 [17], DICOM included the MPEG2 codec that also explores the inter-frame (time) redundancy. One can realize that, in daily practice, the deployment of MPEG based compression on major dynamic imaging modalities is still far from being generalized.

5.3.3 Security and Confidentiality

The private information about citizens' health is highly safeguarded. Medical records contain highly sensitive information about person's health problems, family history, personal behavior and habits. The disclosure of this information can dramatically destroy the social and professional citizens' life.

Citizens are rightly concerned that new health information technologies will make their personal information more accessible and exposed to disclosure or misuse. Their concerns about privacy and confidentiality can only be addressed if they control the process related with their medical records. Current security strategies in healthcare institutions are mostly based on primitive processes such as the simple authentication with username and password. These frequently travel in "clear text" through the network. The "clear text" problem can be quickly eliminated with mechanisms like IPSec or SSL. The authentication trust is still mostly based in the insecure pair of username and password, often predictable. This makes the implementation of effective security services like data privacy, integrity, and non-repudiation of clinic acts/actions impracticable. Security leaks come mostly from the fact that early developments were guided by user patterns in intranets environments. They were not adapted for wider scopes such as the Internet.

A Digital Certificate is an excellent tool for performing identification and authentication in digital transactions. It relies greatly on Public Key

Cryptography (PKC) and establishes a trust relationship between its subject (the owner) and a reliable Certification Authority (CA) that certifies the subjects' identity. One of the most secure and flexible ways to store sensitive information, such as personal details or cryptographic keys, is through the use of smart cards [18,19]. This technology, when correctly combined with emerging technologies, like biometrics, can strongly enforce access control through personal identification and authentication [20].

Relatively to medical image security services, Part 15 (Security and System Management Profiles) of the DICOM standard contemplates several security mechanisms to implement PACS secure environments. Firstly, the norm defines secure exchange of information in the network communications channel or media storage. The DICOM protocol allows a secure transmission by encapsulating the data within a well established network security protocol called TLS (Transport Layer Security). This provides a "secure channel" between communication partners. An alternative is that it also possible to use ISCL (Integrated Secure Communication Layer). Both protocols guarantee authentication, integrity and confidentiality services.

Part 15 of the norm also contemplates the protection of entire DICOM Attributes (that is objects inside one DICOM file) and files, including DICOM directories. This process is done by introducing the files/attributes inside an encrypted envelope CMS (Cryptographic Message Syntax). Another important aspect in the standard is the use of PKC digital signatures to ensure that there is no-repudiation and to ensure the integrity of DICOM objects.

Finally, the use of clinical records in education and research without patient consent requires de-identification. In DICOM structures this can be done by removing Data Elements that contain patient identifying information. For instance, the patient name element with the hexadecimal tag (0010, 0010).

5.3.4 Web-enabled Interfaces

During the last years, the ubiquity of web interfaces have pushed practically all PACS suppliers to develop client applications in which clinical practitioners can receive and analyze medical images, using conventional personal computers and Web browsers. Due to security and performance issues, the utilization of these software packages has been restricted to Intranets [21]. Paradoxically, one of the most important advantages of digital imaging systems is to simplify the widespread sharing and remote access of medical data between healthcare institutions.

The continuous advances of information technologies is motivating new challenges to the healthcare sector. Professionals are demanding new Web-based access platforms for internal work (Intranet). This is since they usually mean negligible installation costs, and are also a way to support Internet based scenarios for remote diagnostic and cooperative work. For this purpose, digital medical images must be integrated with multimedia Web environments.

Several Web-enabled PACS have been deployed but their utilization has been mostly restricted to use on the institutions intranets [21–23].

Traditional PACS components are normally interconnected through a local area network (LAN) that maximize transmission rates, reliability and minimize security risks [24]. Almost all these PACS elements follow the DICOM standard, Part 8 (Network Communication Support for Message Exchange) [25]. This uses communication ports that are typically barred by network administrators in order to avoid intruders and Trojans in their services. This situation severely compromises the development of client/server applications using DICOM. In 2004, the DICOM Standard, Part 18 (Web Access to DICOM Persistent Objects, WADO) [26] was approved. It was conceived to allow web-access to DICOM objects, including images in different formats, utilizing a standardized URL (HTTP-GET).

In addition to the access control problems, distributed PACS face yet another major requirement. For many medical imaging modalities, with increasing volumes of data, the transmission delay over the Internet is normally unacceptable. Doing image compression can be part of the solution but it also raises other problems. This is especially true if there is no other option than the use of severe lossy compression techniques with questionable impacts on image quality and, eventually in diagnosis accuracy.

5.4 HIS and PACS Integration

World globalization causes increased personal mobility. This, when associated with the healthcare services tendency to create specialized diagnostic and therapeutic centres, is creating a higher dispersion of patient clinical records. This is forcing more the healthcare providers to implement telematic infrastructures which should promote the share and remote access to patient clinical data. It is not only the dispersion of data that is increasing, but also the number of diagnostic and routine consultations. The result is that one citizen, during his life, is likely to be associated with a large amount of clinical information that is stored in distinct institutions which are often supported by heterogeneous information systems. There is no efficient means for unique and integrated access to all distributed patient data stored in Healthcare Information Systems (HIS) and within PACS.

The benefits obtained with ready access to all historical patient data are unquestionable to practitioners and to patients. The result is a tremendous improvement in the healthcare services quality, procedural cost-effectiveness, having an immediate benefit to the patients' safety.

The provision of all the types of clinical data in an integrated way is increasing and is a fundamental key for care services in the coming years. A greater objective, such as a truly uniform interface that could provide access to all the historical patient data, must start with small steps over the individual institutions domains and progressively implement and aggregate new items.

A first step, concerning this development model, should be the integration of HIS and PACS in an unique Web interface. This would provide an on-line real time access to authorized healthcare professionals. It would enable the delivery of time-efficient services with enhanced quality. In other words, less time would be used concerning the manual search and integration of possible historic data and more time would be available for providing effective healthcare services.

5.5 Case Study

As mentioned, one of the most important benefits of digital medical imaging and PACS systems is to allow widespread sharing and remote access to medical data, inside or outside institutions. A Web-enabled PACS age emphasized the concept: "Any Image, Anywhere, Any time", promoting an incredible transformation in service delivery across the healthcare institutions. However, in some areas, such as a Cardiology department, their utilization has been mostly restricted to the institution intranet. The difficulties arising in the "real-time medical image remote access" are based on the fact that dynamic imaging modalities (US, XA) impose extra challenges concerning the storage volume required, their management and the network infrastructure needed to support their efficient distribution [21,23]. In general, it is extremely complex to keep "online" in a centralized server the huge volumes of data involved and it is also difficult to retrieve the images in real-time, from outside the institutional broadband network.

This section describes the design and implementation of a cardiac Web-PACS solution that supports a customized DICOM encoding syntax and compression scheme existing in a clinical department. The system includes a telematic platform capable of establishing cooperative telemedicine sessions. It provides a good trade-off between compression ratio and diagnostic quality, low network traffic load, backup facilities and data portability.

5.5.1 Clinical Environment

The main clinical partner for this case was the Cardiology Department of the Hospital of Gaia (CHVNG), in Portugal. Two digital imaging laboratories support this unity. The cathlab (Cardiac Catherization Laboratory) that produces 3000 procedures per year and the echolab (Cardiovascular Laboratory) with 6000 procedures per year [23,27]. A Siemens AXION-Artis cathlab and seven echocardiography machines from various vendors with a standard DICOM output interface, storage-processing server unities, review stations, are the basic PACS entities in this cardiologic department. Recently it has also included a 64 Slice Angio-CT. A schematic overview of the imaging facilities is shown in Fig. 5.2.

The first departmental PACS implementation, which was installed in 1997, allowed the storage of only 12-14 months of angiographic (XA) procedures

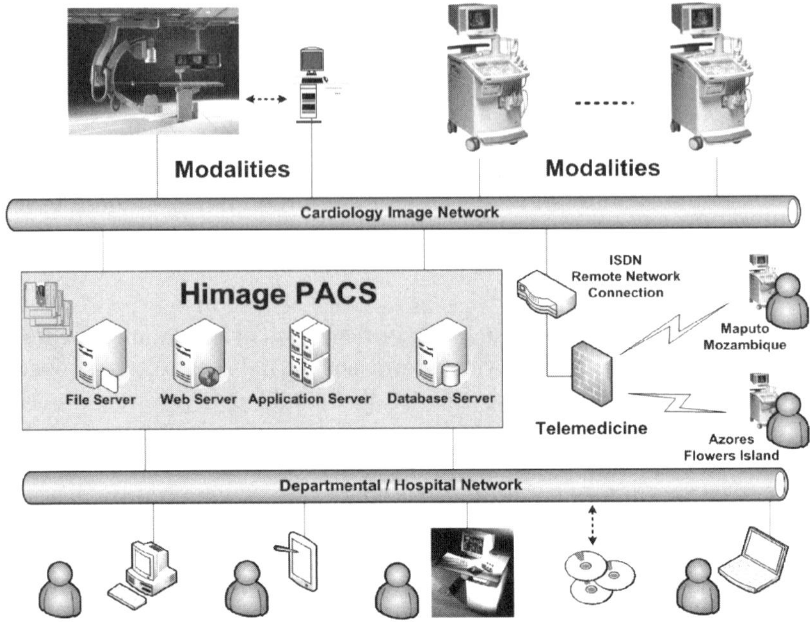

Fig. 5.2. CHVNG Cardiology Infrastructure: Modalities, Network and Systems

based on DICOM JPEG-lossy with a quality factor of 95 and 14-16 months of echocardiography (US) procedures compressed with DICOM JPEG-lossy (quality factor of 85). With this system an US study size can typically vary between 20 and 30 Mbytes and an XA between 40 and 60. This is dependant on the technical characteristics of the equipment, the operator expertise and the procedure type. This volume of information is difficult to cope with if permanent availability and cost-effective transmissions times for remote connections are requirements to be comply. With this framework, transcontinental telemedicine connection was established with the Cardiology Department of the Central Hospital of Maputo (Mozambique) [28].

Focusing on Echocardiographic and Angiographic imaging specifications shows that they are a demanding medical imaging modality when used as digital sources of visual information. The date rate and volume associated with a typical study poses several problems in the design and deployment of systems having acquisition, archiving, processing, visualization and transmission requirements. Digital video compression is a key technology. Given the time-space redundancy that characterizes this type of video signal there are important gains to be made by choosing a compression methodology that copes with both intra-frame and inter-frame redundancies. The definition of an adequate trade-off between the compression factor and the diagnostic quality is a fundamental parameter in the design of both the digital archive and the communications platform.

5.5.2 Compression of Cardiac Temporal Imaging Sequences

In our laboratories, the majority of volume of data produced in the XA and US procedures is related to cine-loop sequences with a significant time-space redundancy, and having a special emphasis in the ultrasound case. Because the JPEG-based DICOM compression algorithms only explore the intra-frame image redundancy, we decided to study the utilization of a more powerful video encoder. This video encoder could also take advantage of the inter-frame redundancy (time). We began our case study with the US modality. After several experiments and trials with other dynamic encoders (MJPEG, MPEG1, MPEG2), the results indicated that the MPEG4 encoding as the best option when considering the suitable trade-offs between image quality and the storage requirements for cardiovascular sequences.

In general, distinct digital video codec's produce significantly different results when used on different medical image sequences types (modalities, cine/still, colour/grayscale, etc). That happens because distinct sequences convey distinct kinds of information. These include still or dynamic, mono-chromatic or polychromatic and always the noise settings are tightly coupled with the clinical acquisition protocol. Our compression trials do not allow us to select the very best among several codec's which could be used for the com-pression of all the types of image/video. From the experience with XA and US cine-loops, we realize that a particular codec may have excellent results only for a limited subset of sequences.

Taking full advantage of object texture, shape coding and from the inter-frame redundancy, our MPEG4 encoding strategy led to results that point to an overall storage boost. The option for 768 kb/s MPEG4 coding pro-duced compressed data sets that are clinically indistinguishable from the large original data sets. The emergence of MPEG4 as a coding standard for the multimedia data and other enhanced encoding facilities appears to be a good alternative for the cost-effective storage and transmission of digital US cine-loops sequences. As a consequence of general satisfaction with the ccholab-MPEG4 choice, we decided to expand these experimentations to the XA sequences. We made an evaluation test with MPEG4 and other dynamic encoders more oriented to grayscale modalities (8bits/pixel).

5.5.2.1 Data Volume Analysis

More than 260000 US studies have been done so far and stored in this Car-diac PACS. For example, a typical Doppler color run (RGB) with an opti-mized time-acquisition (15–30 frames) and a sampling matrix (480×512), rarely exceed 200–300 kB. As shown in Fig. 5.3, typical compression ratios can be from 35:1 for a single cardiac cycle sequence to 100:1 in multi-cycle sequences.

With these averaged figures, even for an examination intensive laboratory, it is possible to have all historic procedures online or distribute them with

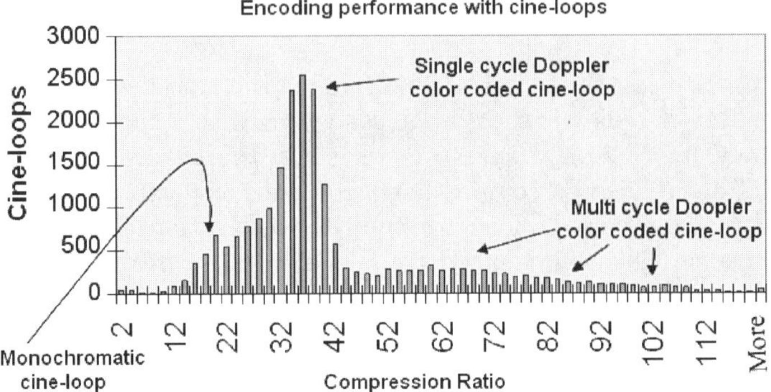

Fig. 5.3. Compression Performance

Table 5.2. Himage PACS – data volume statistics

Number of Procedures (Examinations)	15.921
N° of Images (Still + Cine-loops)	325.499
Total of Space Volume	65.936.030.821 Bytes (61,4GB)
Average Procedure Size	4.044 Kbytes
Average File Size	198 Kbytes
Average Files / Procedure	20 (mostly of cine-loops)

reduced transfer time over the network. This is a critical issue when dealing
with costly or low bandwidth connections. According to the results of the
user's assessment, the achieved compression ratios, do not compromise diagnostic quality [23]. They reduce significantly the download time and display
latency of images. It is possible to claim that the usual near-line and off-line
storage requirements are more a matter of backup policy than a mandatory
component for the daily working routine of these laboratories.

In Table 5.2 we give a set of global statistics obtained from the first 15921
studies. Concerning the cathlab, the average cine-loop size for XA grayscale
images (8bits/pixel) with a 512x512 matrix and 90 frames, rarely exceed 200-
300kB (compression 80:1).

For the previous values, it is possible to have all historic procedures online
or to distribute them with reduced latency. This is a critical issue when dealing
with costly or low bandwidth connections such as telemedicine or telework.
The new encoding image syntax and compression format simplify the integration with the HIS information to be displayed in the Web environment.

5.5.2.2 Image Quality Evaluation

A clinical evaluation was done for assessing the MPEG4 (768kbps) encoded
sequences in DICOM cardiovascular ultrasound images. It was conducted by
the physicians of the Cardiology Echolab who have blindly compared the

compressed images against the uncompressed originals. Qualitative and quantitative results have been observed [23]. An interesting fact from this study was that, in a simultaneous and blind display of the original against the compressed cine-loops, 37% of the trials have selected the compressed sequence as the best image. This suggests that other factors related to the viewing conditions are more likely to influence the observer performance than the image compression itself.

The quantitative and qualitative assessments of the compressed images show that the quality is retained without impairing the diagnostic value. Using compression factors of the same magnitude in other DICOM coding standards JPEG [29] MPEG1 [3] leads, on these cases, to a severe decrease on image quality.

5.5.3 DICOM Private Transfer Syntax

The actual Internet image/video compression market is pushing the development of new and better codecs. Here, we believe that it will be very difficult for the DICOM Standard to follow this evolution. For instance, the MPEG4 proved to be a good solution for US in general. However, the DICOM standard has only recently (2004) adopted MPEG2 as the codec able to explore the inter-frame redundancy [17].

Based on the previous evidences and aiming to ensure more flexibility, it was decided not to insert the MPEG4 directly in the TLV (Tag Length Value) DICOM data structure [1]. That is, to solve directly the US "specific problem". Instead, pointed out in Fig. 5.4, a multimedia container that dynamically supports different encoders was developed. The container has a simple structure, including a field to store the encoder ID code. This is the same as that used by the AVI RIFF headers [30]. When it is necessary to decompress the images, the Himage-PACS solicits the respective decoder service to the operating system, as in any other multimedia application.

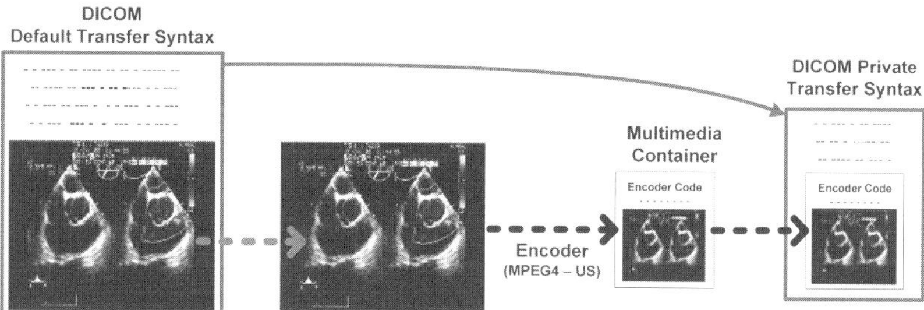

Fig. 5.4. Conversion from the DICOM "default transfer syntax" into the "private syntax"

Tag	VR	Contents	length	Comments	
....					
(0002,0010)	UI	[1.2.840.10008.1.2]	# 18,	DefaultTransferSyntaxUID	(before)
(0002,0010)	UI	[1.2.826.0.1.3680043.2.682.1.4]	# 22,	PrivateTransferSyntaxUID	(after)
....					
....					
(7fe0,0010)	OW	0000\0000\0000\......	# 33177600,	PixelData	(before)
(7fe0,0010)	OW	4952\4646\355c\0004\...	# 275968,	PixelData	(after)

Fig. 5.5. DICOM dump: DICOM "default transfer syntax" is compared with the "private syntax"

This approach is an optimized high level software solution. If a more efficient codec is available in the future, we only need to change a single parameter in the "Himage Conversion Engine".

In Fig. 5.5 parts of the DICOM objects representation for the "default transfer syntax" are compared with the "private syntax". In this particular case a US 25 frames image sequence with a RGB 576×768 matrix. It is possible to observe two important aspects. Firstly, the DICOM *DefaultTransferSyntaxUID* identifier (1.2.840.10008.1.2) is replaced by a private *PrivateTransferSyntaxUID* (1.2.826.0.1.3680043.2.682.1.4). This later UID is based on the root UID 1.2.826.0.1.3680043.2.682 and was requested by our workgroup. Secondly, the "PixelData" field size "is reduced" by 120 times (33177600/275968).

5.5.4 Himage PACS Solution

In 2003, we finished the second generation of the Himage PACS [23]. The JPEG DICOM compressing algorithm was replaced by a more powerful video codec container. This is capable of coping better with intra-frame and inter-frame redundancies. The distinct modules composing the Himage system included the acquisition of uncompressed the DICOM (default transfer syntax) sent by modality machines. The processing and creation of DICOM private syntax files and the subsequent storage of the patient study in the online server. This information includes alphanumeric element data and images. The original images (that is JPEG lossless to grayscale modalities) are saved and kept online during six months. The DICOM private syntax sequences are available during the system lifetime. The developed client application module includes a DICOM viewer that allows the clinical staff to visualize standard DICOM and also private syntax exams that are available in the Himage database.

5.5.4.1 Web-enabled Interface

As stated, the lack of performance in studies transfer has restricted the use of PACS Web-enabled solutions to Intranets. This is especially true in dynamic imaging modalities. The successful results obtained using imaging studies data volumes with the private transfer motivated us to undertake the development a Himage Web version [21]. Here the new software goal was to provide the physicians with a Web-based PACS that is easily accessible using a web-browser without a complex local installation.

Web solutions can be done with different technology architectures [31]. In the case of PACS the graphical user interface is extremely demanding. Our approach is based on the .NET framework which allows a smooth integration with existing code. All the core functions to manipulate DICOM structures and images pixel data were written in C++ and made available as DLL components. The graphical components were developed in Visual Basic (OCX) and directly embedded in the ASPX .NET pages. This ActiveX viewer (i.e. the OCX) allows the direct integration of private DICOM files with the Web contents. The communication between the dynamic HTML contents and the OCX binary is done in JavaScript. The images retrieve is supported by a webservice developed in C#.

The Web-enabled PACS interface is fully operational and provides all previous functionalities and has the same performance and flexibility as the previous Intranet "winforms" version. The application setup is very simple. It is downloaded from the web server and is automatically installed in the first Himage utilization, after user authorization. The communication between the Web server and the client browser is implemented in the same way as other web service and it is encrypted with HTTPS.

Graphically, the Himage main window includes a grid box with the patients list and a cine preview of the three sequences for the current selection (Fig. 5.6). The user can "search" procedures using different criteria: patient name, patient ID, procedure ID/type/date, source institutions and equipment. A second graphical application layer provides interfaces to the different modules: communications to telecardiology, DICOM viewer (Fig. 5.7 and Fig. 5.8), report (Fig. 5.9) and export.

In the Himage DICOM viewer window (Fig. 5.7 and Fig. 5.8) it is possible to visualize the image sequences of still and cine-loops, select frames from various sequences that are sent to the report area. Other traditional functionalities have been included like, for example, the image manipulation tools (contrast/brightness), the printing capacities and the export of images in distinct formats (DICOM3.0 default transfer syntax, AVI, BMP, JPEG). It is also possible to copy a specific image to the clipboard and paste it into some other external application.

The viewer window, in the compare mode, can work in either of two ways: Automatic or Manual. The automatic mode is used for a specific type of procedure. The "stress eco" (Fig. 5.8) is based on the information stored in the DICOM file headers. The Himage makes a simultaneous and synchronized display of the different "Stages" of a heart "View". The term "Stage" is defined as a phase in the stress echo exam protocol. The "View" is the result of a particular combination of the transducer position and orientation at the time of the image acquisition. The manual mode allows the user a selection

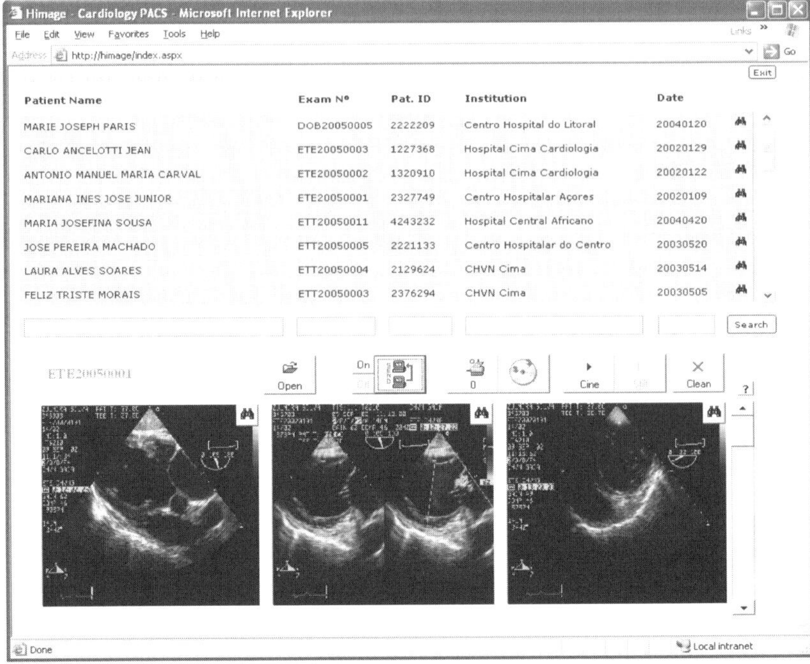

Fig. 5.6. Web Himage – Main window

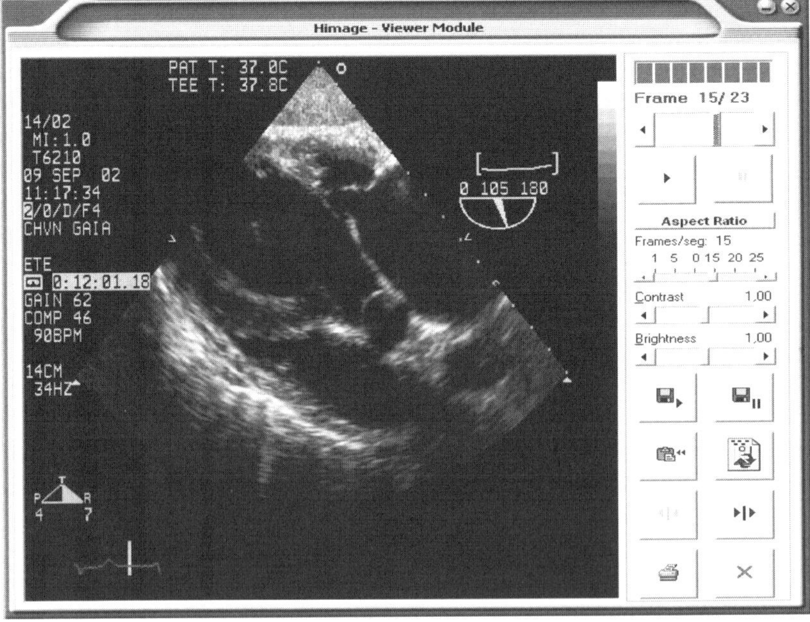

Fig. 5.7. Web Himage – Viewer window in the normal mode

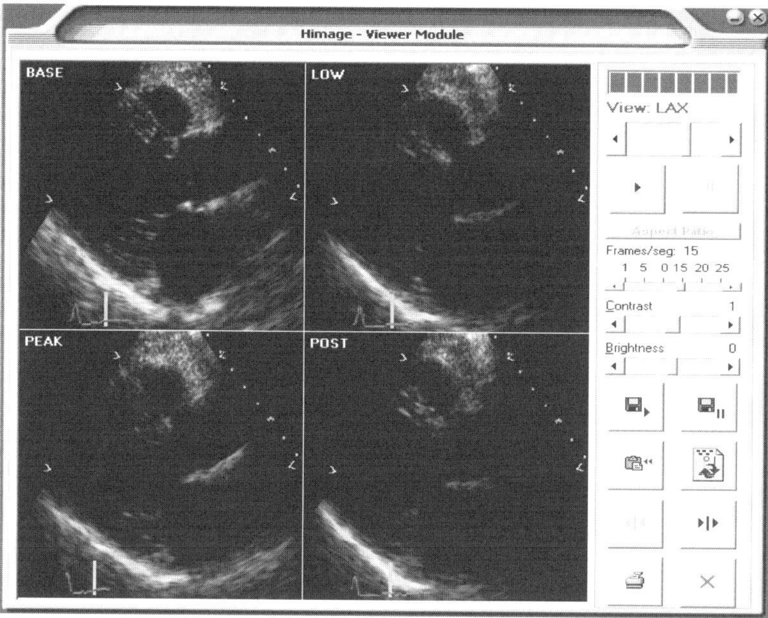

Fig. 5.8. Web Himage – Viewer window using the compare mode

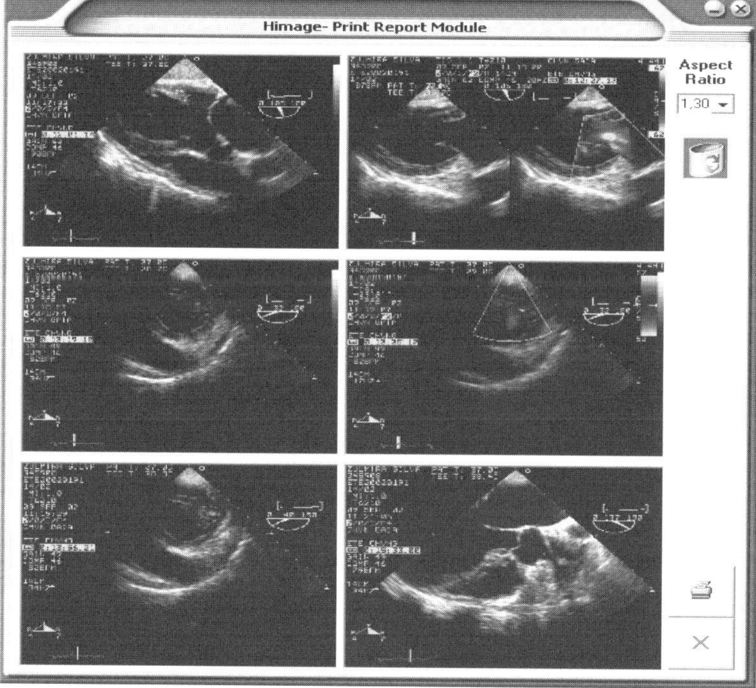

Fig. 5.9. Web Himage – Report module

of distinct images of the same and/or of different procedures that can be displayed in a defined window matrix.

In the report module (Fig. 5.9), the user can arrange the images location or delete some frames using a drag-and-drop functionality. Finally, the output images matrix (2×3 or 3×3) is used with the clinician report to generate a RTF file with is compatible with any common text editor. The user can customize the base template used to generate the report file. This would probably contain institution logo and report headers.

The export module allows the procedure to be recorded in an external storage device, such as a CDROM or DVD-ROM, using the uncompressed DICOM default transfer syntax format or the AVI format. The media storage unit also has a standalone DICOM viewer application. This automatically runs when the disk is introduced into the PC drive.

We have permanently available though this system more than 26000 US procedures.

5.5.4.2 Telematic Platform

The huge volumes of cardiac US medical image data are not easy to transfer in a time and cost-effective way. Healthcare professionals will not use telemedicine or telework platforms if they must wait 1–3 hours to download or upload a clinical imaging study with a diagnostic quality. The reduced image size of Himage US procedures is a factor which will make the acceptance of the technique more feasible for new clinical telematic applications. This is especially important in environments with limited bandwidth or a reduced communications budget. The Himage includes a communication platform that allows the creation of customized study bundles, where there is more than one study, to be sent to predefined remote institutions. With this facility the physician can select images sequences, make annotations, append extra comments and send the examination over the network to the remote unit. The main Himage Web interface (Fig. 5.6) has a group of buttons "on/off" and "send" that allow the activation and the selection of sequences which are to be sent to the communication module. The user just needs to choose the target institution and send the data using an email paradigm.

The first telemedicine project has been established with Cardiology Department of the Central Hospital of Maputo (Mozambique) [28]. Both clinical partners are equipped with echocardiography machines embedded with standard DICOM output interfaces, videoconference platforms and a network infrastructure. They have also been installed two server computer to support the PACS. The Portuguese medical partner provided the clinical training and maintains remote support for the diagnostics and therapeutic decisions, in offline and real-time sessions [32]. Recently, we started a new project unity at Flores Island in the Azores archipelago.

Telemedicine projects are, at present, exclusively supporting ultrasound transmission. The sessions can be conference-based teleconsultations for analysis of real-time clinical cases, or the analysis of previously transferred exam files (DICOM private syntax). In this last case the remote physician can select the desired echo images sequences, add annotations, append extra comments and send the information to the Gaia Hospital. This information is stored in one dedicated server and is accessible by the Gaia cardiovascular ultrasound specialists. Both methods can be used simultaneously. The physician in Gaia can open and examine a clinic record file previously sent from the remote area, and simultaneously use the videoconference equipment for face-to-face consultation and for therapy decision. The communication relies on n×64 kb/s ISDN channels. A complete US exam typically takes 2–5 min using a 64 kb channel.

The described cardiovascular telemedicine task, using the referred clinical modus operandi and having limited telecommunications resources available, was impracticable until the development of the Himage system using DICOM private syntax.

5.5.4.3 HIS-PACS Integration

Our development efforts on HIS-PACS integration resulted in a Web interface that combines the patient information retrieved from the HIS system with PACS images [33]. This makes the alphanumeric and multimedia data available in an Internet browser (Fig. 5.10).

The DICOM private transfer syntax, which can support any video encoder installed on a Windows-based station, allows portable and easy integration of medical images in Web environments. The reduced file size is extremely important to provide time-efficient downloads. A software module was developed, called the "Himage Integration Engine". It features two integration options: The first provides a Web binary (ActiveX) viewer that allows the direct integration of the private DICOM files in the Web contents. Secondly, the engine can extract the encoded image data from the DICOM and encapsulate it using an AVI file format for insertion into a Web document.

5.6 Advanced Applications

It is now recognized that the application potential of a generic PACS may far exceed the long standing concepts of archival and communication of imaging information. The emergence of PACS based medical imaging informatics infrastructures [34, 35] is driven by several applications that either depend on or do benefit from the seamless access to image databases and related metadata. Image Fusion based studies, Computer-Aided Diagnosis, Computer-Assisted Surgery, Image based data mining, e-Learning are some prominent application areas that are broadening the original PACS frontiers towards the decision support perspective either in enterprise-wide contexts or in departmental contexts. Cardiovascular imaging is at the forefront of these trends and,

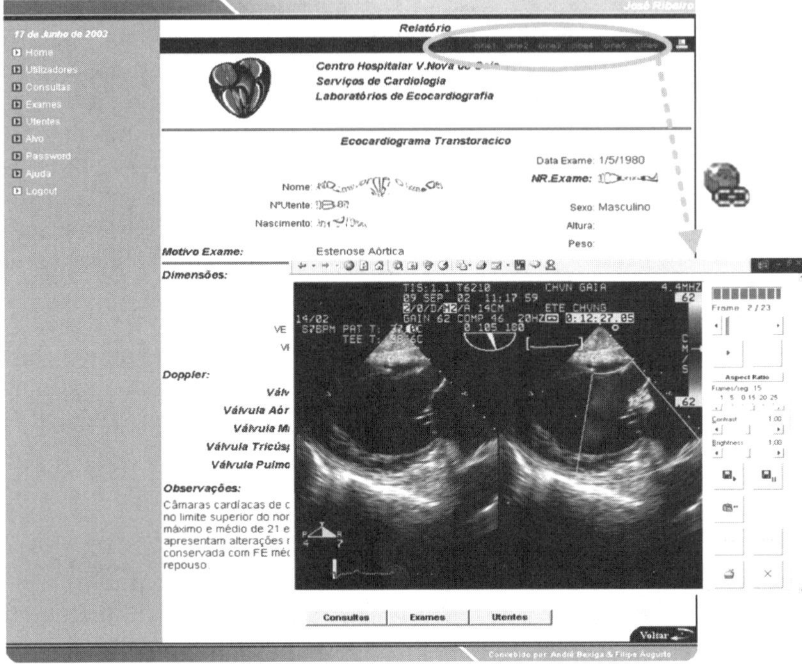

Fig. 5.10. Integration of the HIS and the Himage PACS in a web interface

within this context, we decided to prospectively elaborate on the following
PACS based applications: study co-registration and content based retrieval.

5.6.1 Study Co-registration

The literature is filled with multi-modality based studies and there is a well
established framework of algorithms able to perform image co-registration.
For example, the survey references [36–38] show that the vast majority of
methods and applications are still based on a static approach where modality
fusion is carried on a stop-action basis.

Motion dominates heart physiology and, as can be expected, co-
registration methods in cardiac imaging often have to cope with this major
factor. Within the cardio-thoracic setting, the number of potential external
or anatomical landmarks, that could be used to facilitate the process of
study co-registration, is remarkably fewer than those encountered in other
anatomical domains such as the brain.

Another important trend in this particular field is the steadily growing
acceptance of tomographic modalities such as MRI and, particularly, fast
MDCT. Here new clinical evaluation paradigms are being driven by the ability of nearly stop-action imaging of cardiac volumetric datasets where most

Table 5.3. Co-registration frameworks in cardiac imaging

	Single Modality	Multi Modality
Intra-subject	Registration of temporal sequences Temporal deformation of anatomical. structures (heart, chest, blood flow) Study comparison of Pre and Pos interventional procedures Growth, Pathology follow-up	Myocardial viability Perfusion. MRI, PET, US Metabolism PET, SPECT Contractile function MRI, US, CT Interventional Imaging registration of pre- and intra-operative images MRI, US
Inter-subject	Model-based segmentation Building of digital atlases Registration/matching with an anatomical atlas Spatial normalization, study of anatomical variability	Anatomo-functional normalization

of motion artifacts are of negligible effect. The co-registration domain is now truly 4D and the computational burden gains relevance since multislice – multiphase acquisitions may boost the volume of data well beyond 1GB per procedure.

Generally image fusion studies may occur in several instances dictated mainly by their clinical interest. Table 5.3 summarizes the common frameworks where cardiac image fusion has shown potential usefulness.

Accordingly to the taxonomy of [39] one can encounter co-registration methods that are either driven by matching geometric image features or by exploring voxel similarity measures. In any case the actual mathematical mapping associated with the particular co-registration process may be classified as rigid, elastic or affine. See [39] for a survey of methods and an extensive list of references to applications. As a recent paradigmatic application of intra-subject – intra-modality see for example [40] where pre and post stress echocardiographic sequences are mapped. The emergence of fast MDCT has been providing new challenges for co-registration [41]. The superposition of planar coronary angiography and heart volumes provided MDCT has also been reported as an auxiliary tool to assist the Percutaneous Transluminal Coronary Angioplasty (PTCA) procedure itself.

The interest for appropriately co-registered multi-modal information has by far surpassed the algorithmic questions and the major proof is grounded on the emergent hybrid systems like PET-CT [42, 43] or SPECT-CT [44]. DICOM objects for these newcomers are under development as necessary for smooth PACS integration.

Co-registration theory and methods are, in fact, quite orthogonal to PACS concepts however, since the co-registration studies are performed on a routine basis the advantage of the PACS infrastructure becomes clearly visible.

Problems such as workflow, data volumes, software management for co-registration and visualization would require outstanding efforts without the PACS functionalities.

Moreover, this necessity for the integration of visual information may sustain the concept of an "Organ" oriented PACS against the more traditional view of a PACS confined to Radiology.

5.6.2 Content-based Access

Content-Based Image Retrieval (CBIR) has also been a demanding research topic. The medical imaging field appeared as a major application scenario for content-based access as a support tool in the clinical decision-making process. The integration of these access methods into current PACS software architectures, although not a breakthrough novelty, is still in its infancy. There are many problems that remain to be solved before it can be practical and generally deployed. PACS systems are still mainly regarded as a repository instance and its search functions are mostly determined by the patient or the imaging study characteristics, available as structured textual information within the DICOM files.

Any attempt to reengineer this crude repository approach into next generation of medical imaging retrieval systems will have to consider the storage subsystem of a PACS into a much broader sense. Besides the storage component itself, a successful PACS based CBIR will depend on new indexing schemes which encompass textual and visual tokens. Retrieving image data from large databases will depend both on the nature of these tokens and how well they may be considered as representatives of a particular medical imaging population either as normality or as pathological indicators. See [45] for an overview of the distinguishing factors that may be used to describe information in medical imaging frameworks.

Developing appropriate visual queries is at the very centre of research in CBIR. The vast majority of implemented systems capture low-level features associated with graylevel/color, texture, shape and which may be assembled either in a global or in a local mode. These features are then directly used as visual queries or are passed to higher level image processing steps. These will hopefully, enrich the semantic contents of the visual query. The general trend is that every processing effort will be worthwhile given its capabilities to shorten the so called semantic gap. This often used term in the CBIR jargon, basically means the loss of information associated with the inevitable step of data reduction when moving from the image domain to the feature domain.

What is the best feature space? What is the appropriate similarity measure at retrieval time? How domain specific knowledge may be integrated in order to shorten the search space? These are general and ever open questions for which we can find practical answers only in rather focused scenarios. See for example [46–50] as implementation cases where quite different feature spaces and retrieval strategies were used.

So far, instead of pursuing the rather intangible goal of pure semantic CBIR most approaches have implemented retrieval engines based on hybrid querying. Most PACS databases are easily linked with textual patient records or, even better, diagnosis related information is available in DICOM structured reporting [51–53]. Hybrid querying can boost retrieval performance partly because of the well established technology of text based retrieval.

Cardiac imaging, as mentioned before, poses its own set of challenges at each of the possible image processing instances. Cardiac CBIR is a tremendous challenge if one considers all the potential dimensions involved.

Looking at the traditional dynamic modalities of XA and US in the context of CBIR prompt us for the more appropriate domain of Content Based Video Retrieval (CBVR). There is abundant literature in the area of CBVR in multimedia environments but, up to the author's knowledge, its applications in the medical domain are scarce. Consider for example the US case where the feature space must include the temporal dimension. The general issue of the appropriateness of the feature space must be considered. Can we use Principal Component Analysis, (PCA) or Independent Component Analysis (ICA) as general tools for data reduction? Is there such a thing as a set of eigen cineloops that could be representatives of valve pathologies? What kind of similarity metrics should be used? These are some of the many open questions which require debate and research.

The 4D modalities such Cine-MRI and the more recent MDCT are also profuse sources of research problems concerning CBIR. These include matters such as what's the most effective way of querying the database? Shall we rely on a set of slice based features or on representative volumetric rendering?

Despite the impressive tasks of overcoming these technical difficulties, the potential benefits of CBIR in medical imaging are readily recognized. Current literature points out several application domains where integration of CBIR will be of major benefit [54]. Examples are in the areas of teaching, research and diagnosis. From the management perspective, automated annotation and coding is another topic that will likely benefit from appropriate and validated CBIR systems.

Availability of large imaging repositories is obviously a significant advantage for lecturing purposes. Here, patient demographic information is of marginal interest when compared with the need to access a set of images that may be regarded as a typical representative of a given disease. Moreover an ideal CBIR system must cope with situations, which are not rare, where visually similar cases may be associated with different diagnosis. We now have many Internet based sites offering teaching files with common browsing capabilities. A very few of these are supported on a PACS infrastructure and CBIR is normally absent. Turning a PACS into an academic tool is an evolving trend and in some cases [55] it becomes the preferred electronic presentation platform for multidimensional cases such as the recent 4D cardiac studies. The extension of PACS access to the Internet is the basis for modern e-Learning. Moreover, CBIR integration will mean a significant step in active learning.

Accessing a large scale image database with visual queries may also contribute for finding new relationships between visual appearance and disease manifestations. The contrary may also be a matter of research interest if the scope of visual variations should be addressed when studying a particular disease. Large scale studies driven by image mining are much more likely to point out new relationships between imaging patterns and disease manifestations.

Looking at CBIR as a CAD tool itself leads us to the hardest and most noble application instance of this kind of image retrieval. The general approach is to provide the clinical decision maker with as much integrated information as possible in the shortest time window. Visual querying may act in a dichotomic way either providing cases through similarity or by dissimilarity searches if some kind of distance to normality is best suited to clinical reasoning.

Whatever the approach, much has to be done before wide acceptance of this kind of CAD tool by the clinical community. Retrieval algorithms must be thoroughly tested over reference and publicly available databases, their impact on diagnosis performance must be assessed [56–58] and finally integration on clinical routine must be promoted though appropriate user-interfaces.

Every clinical department has to deal with annotation and coding procedures which are often time consuming and require skilful human resources. CBIR systems in imaging departments may speed-up the coding tasks since similar previously annotated cases are potentially available for comparison. It is possible to devise software tools that combine CBIR output with available coding guidelines in order to improve the overall quality and reproducibility of the coding tasks.

Departmental clinical quality assurance is another important management issue that may be facilitated by PACS based CBIR systems. Diagnosis consistency may be assessed and image quality may be confronted with acquisition protocols which are now fully described by the DICOM headers.

5.7 Conclusions

The dissemination of digital medical imaging systems has been steadily increasing in healthcare centers, representing now one of the most valuable tools supporting medical decision and treatment procedures. Until recently, the utilization of this digital technology was confined to the dispersed equipment and the number of successful cases that use centralized digital archives was reduced. The provision of a PACS increases practitioner's satisfaction as a result of the faster and ubiquitous access to image data. In addition, it reduces the logistic costs associated with the storage and management of image data.

Digital medical images have opened the door to a large variety of storage and processing techniques which allow the extraction of more and better information from the acquired data. Current clinical practice provides many examples of image processing applications that rely on PACS resources. These processes are likely to benefit from the current research on modern indexing and retrieval strategies.

In this chapter we have highlighted past, current and expected major problems, standards, and solutions related to the use of PACS. We have presented the Himage PACS that was the core infrastructure to the implementation of the first fully digital Echocardiography Laboratory in Portugal. This PACS applies lossy compression techniques to dynamic images as a way to explore spatial-time redundancy. The compression and storage results were successful assessed by clinicians and present on-line the entire patients' history. This characteristic has proved to be particularly important in telemedicine scenarios using a low communication bandwidth. The framework also provides a web interface supported by a powerful access control mechanism.

References

1. DICOM-P5, Digital Imaging and Communications in Medicine (DICOM), Part 5: Data Structures and Encoding. 2004, National Electrical Manufacturers Association

2. DICOM-P6, Digital Imaging and Communications in Medicine (DICOM), Part 6: Data Dictionary. 2000, National Electrical Manufacturers Association

3. Thomas, J., et al., Digital Compression of Echocardiograms : Impact on Quantitative Interpretation of Color Doppler Velocity. Digital Cardiac Imaging in the 21st Century: A Primer, 1997

4. Kennedy, J.M., et al., High-speed lossless compression for angiography image sequences. Proceeding of SPIE - Medical Imaging 2001: Visualization, Display, and Image-Guided Procedures, Seong K. Mun; Ed, 2001. 4319: pp. 582–589

5. Huang, H.K., PACS and Imaging Informatics: Basic Principles and Applications. 2004, New Jersey: Wiley-Liss; 2nd edition

6. Chen, T.J., et al., Quality degradation in lossy wavelet image compression. J Digit Imaging, 2003. 16(2): pp. 210–5

7. Maede, A. and M. Deriche, Establishing perceptual limits for medical image compression. Proceedings of SPIE - Image Perception and Performance, 2001. 4324: pp. 204–210

8. Brennecke, R.u., et al., American College of Cardiology/ European Society of Cardiology international study of angiographic data compression phase III: Measurement of image quality differences at varying levels of data compression. J Am Coll Cardiol, 2000. 35(5): pp. 1388–1397

9. Kerensky, R.A., et al., American College of Cardiology/European Society of Cardiology international study of angiographic data compression phase I: The effects of lossy data compression on recognition of diagnostic features in digital coronary angiography. J Am Coll Cardiol, 2000. 35(5): pp. 1370–1379

10. Tuinenburg, J.C., et al., American College of Cardiology/ European Society of Cardiology international study of angiographic data compression phase II: The effects of varying JPEG data compression levels on the quantitative assessment of the degree of stenosis in digital coronary angiography. J Am Coll Cardiol, 2000. 35(5): pp. 1380–1387

11. Umeda, A., et al., A low-cost digital filing system for echocardiography data with MPEG4 compression and its application to remote diagnosis. Journal of the American Society of Echocardiography, 2004. 17(12): pp. 1297–1303
12. Segar, D.S., et al., A Comparison of the Interpretation of Digitized and Video-tape Recorded Echocardiograms, Journal of the American Society of Echocardiography, 1999. 12(9): pp. 714–719
13. Karson, T.H., et al., Digital storage of echocardiograms offers superior image quality to analog storage, even with 20:1 digital compression: Results of the digital echo record access study. Journal of the American Society of Echocardiography, 1996. 9(6): pp. 769–778
14. Huffman, D.A., A method for the construction of minimum redundancy codes. Proceedings IRE, 1952. 40: pp. 1098–1101
15. DICOM-SUPL61, Digital Imaging and Communications in Medicine (DICOM), Supplement 61: JPEG 2000 Transfer Syntaxes. 2002, National Electrical Manufacturers Association
16. JPEG2000, ISO/IEC JTC1/SC29 WG1, FCD15444: JPEG 2000 Image Coding System Part 1 - Final Committee Draft Version 1.0. 2000
17. DICOM-SUPL42, Digital Imaging and Communications in Medicine (DICOM), Supplement 42: MPEG2 Transfer Syntax. 2004, National Electrical Manufacturers Association
18. Marvie, R., Pellegrini, M. et al. Value-added Services: How to Benefit from Smart Cards. in GDC2000. 2000. Montpellier, France
19. Gobioff, H., et al. Smart Cards In Hostile Environments. in Proceedings of The Second USENIX Workshop on Electronic Commerce. 1996. Oakland, U.S.A.
20. Hachez, G., F. Koeune, and J. Quisquater, Biometrics, Access Control, Smart Cards: A Not So Simple Combination, in Security Focus Magazine. 2001 October
21. Costa, C.M.A., et al., A demanding web-based PACS supported by web services technology. Medical Imaging 2006: Visualization, Image-Guided Procedures, and Display. Edited by Cleary, Kevin R.; Galloway, Robert L., Jr. Proceedings of the SPIE, 2006. 6145: pp. 84–92
22. Costa, C., et al., A New Concept for an Integrated Healthcare Access Model. MIE2003: Health Technology and Informatics - IOS Press, 2003. 95: pp. 101–106
23. Costa, C., et al. Himage PACS: A New Approach to Storage, Integration and Distribution of Cardiologic Images. in PACS and Imaging Informatics - Proceedings of SPIE. 2004. San Diego - CA - USA.
24. Bernarding, J., A. Thiel, and T. Tolxdorff, Realization of security concepts for DICOM-based distributed medical services. Methods Inf Med, 2000. 39(4–5): pp. 348–52
25. DICOM-P8, Digital Imaging and Communications in Medicine (DICOM), Part 8: Network Communication Support for Message Exchange. 2003, National Electrical Manufacturers Association
26. DICOM-P18, Digital Imaging and Communications in Medicine (DICOM), Part 18: Web Access to DICOM Persistent Objects (WADO). 2004, National Electrical Manufacturers Association
27. Silva, A., et al. A Cardiology Oriented PACS. in Proceedings of SPIE: Medical Imaging. 1998. San Diego - USA.
28. Costa, C., et al., A Transcontinental Telemedicine Platform for Cardiovascular Ultrasound. Technology and Health Care - IOS Press, 2002. 10(6): pp. 460–462

29. Karson, T., et al., JPEG Compression Echocardiographic Images: Impact on Image Quality. Journal of the American Society of Echocardiography, 1995. Volume 8 - Number 3

30. IBM Corporation and Microsoft Corporation, Multimedia Programming Interface and Data Specifications 1.0. 2001

31. Zhanga, J., J. Suna, and J.N. Stahl, PACS and Web-based image distribution and display. Computerized Medical Imaging and Graphics - Elsevier, 2003. 27: pp. 197–206

32. Costa, C., et al. Arquivo e transmissão digital de imagem para tele-ecocardiografia transcontinental. in II Congresso Luso Moçambicano de Engenharia. 2001. Maputo - Moçambique

33. Costa, C., J.L. Oliveira, and A. Silva. An Integrated access interface to multimedia EPR. in CARS2003. 2003. London - UK

34. Huang, H.K., Medical imaging informatics research and development trends–an editorial. Computerized Medical Imaging and Graphics, 2005. 29(2-3): pp. 91–93

35. Huang, H., PACS and Imaging Informatics. 2004, New Jersey: Wiley-Liss

36. Hill01, et al., Medical image registration. Physics in Medicine and Biology, 2001(3): pp. 1–45

37. Fitzpatrick, J.M., D.L.G. Hill, and J. Calvin R. Maurer, Image Registration, in Handbook of Medical Imaging, M. Sonka and J.M. Fitzpatrick, Editors. 2000, SPIE Press. pp. 447–514

38. Maintz, J.B.A. and M.A. Viergever, A survey of medical image registration. Medical Image Analysis, 1998. 2(1): pp. 1–36

39. Makela, T., et al., A review of cardiac image registration methods. Medical Imaging, IEEE Transactions on, 2002. 21(9): pp. 1011–1021

40. Shekkar04, et al., Registration of real-time 3-D ultrasound images of the heart for novel 3-D stress echocardiography. Medical Imaging, IEEE Transactions on, 2004. 23(9): pp. 1141–1149

41. Lacomis, J.M., et al., Multi-Detector Row CT of the Left Atrium and Pulmonary Veins before Radio-frequency Catheter Ablation for Atrial Fibrillation. Radiographics, 2003. 23(90001): pp. 35S–48

42. Kapoor, V., B.M. McCook, and F.S. Torok, An Introduction to PET-CT Imaging. Radiographics, 2004. 24(2): pp. 523–543

43. Schwaiger, M., S. Ziegler, and S.G. Nekolla, PET/CT: Challenge for Nuclear Cardiology. J Nucl Med, 2005. 46(10): pp. 1664–1678

44. Utsunomiya, D., et al., Object-specific Attenuation Correction at SPECT/CT in Thorax: Optimization of Respiratory Protocol for Image Registration, Radiology, 2005. 237(2): pp. 662–669

45. Tagare, H.D., C.C. Jaffe, and J. Duncan, Medical Image Databases: A Content-based Retrieval Approach. J Am Med Inform Assoc, 1997. 4(3): pp. 184–198

46. Aisen, A.M., et al., Automated Storage and Retrieval of Thin-Section CT Images to Assist Diagnosis: System Description and Preliminary Assessment. Radiology, 2003. 228(1): pp. 265–270

47. Shyu, C.-R., et al., ASSERT: A Physician-in-the-Loop Content-Based Retrieval System for HRCT Image Databases. Computer Vision and Image Understanding, 1999. 75(1–2): pp. 111–132

48. Sinha, U. and H. Kangarloo, Principal Component Analysis for Content-based Image Retrieval. Radiographics, 2002. 22(5): pp. 1271–1289

49. Bucci, G., S. Cagnoni, and R. De Dominicis, Integrating content-based retrieval in a medical image reference database. Computerized Medical Imaging and Graphics, 1996. 20(4): pp. 231–241
50. Robinson, G.P., et al., Medical image collection indexing: Shape-based retrieval using KD-trees. Computerized Medical Imaging and Graphics Medical Image Databases, 1996. 20(4): pp. 209–217
51. Taira, R.K., S.G. Soderland, and R.M. Jakobovits, Automatic Structuring of Radiology Free-Text Reports. Radiographics, 2001. 21(1): pp. 237–245
52. Hussein, R., et al., DICOM Structured Reporting: Part 1. Overview and Characteristics. Radiographics, 2004. 24(3): pp. 891–896
53. Hussein, R., et al., DICOM Structured Reporting: Part 2. Problems and Challenges in Implementation for PACS Workstations. Radiographics, 2004. 24(3): pp. 897–909
54. Muller, H., et al., A review of content-based image retrieval systems in medical applications–clinical benefits and future directions. International Journal of Medical Informatics, 2004. 73(1): pp. 1–23
55. Sigal, R., PACS as an e-academic tool. International Congress Series, 2005. 1281: pp. 900–904
56. Obuchowski, N.A., ROC Analysis. Am. J. Roentgenol., 2005. 184(2): pp. 364–372
57. Blackmore, C.C., The Challenge of Clinical Radiology Research. Am. J. Roentgenol., 2001. 176(2): pp. 327–331
58. Weinstein, S., N.A. Obuchowski, and M.L. Lieber, Clinical Evaluation of Diagnostic Tests. Am. J. Roentgenol., 2005. 184(1): pp. 14–19

Attacking the Inverse Electromagnetic Problem of the Heart with Computationally Compatible Anatomical and Histological Knowledge

Efstratios K Theofilogiannakos[1], Antonia Anogeianaki[2,3], Anelia Klisarova[2], Negrin Negrev[3], Apostolos Hatzitolios[4], Petros G Danias[5,6], and George Anogianakis[1]

[1] Department of Physiology, Faculty of Medicine, Aristotle University of Thessaloniki, Greece
[2] Department of Imaging Diagnostics and Nuclear Medicine, Medical University of Varna, Bulgaria
[3] Department of Physiology, Medical University of Varna, Bulgaria
[4] Department of Internal Medicine, AHEPA Hospital, Faculty of Medicine, Aristotle University of Thessaloniki, Greece
[5] Hygeia hospital, Athens, Greece
[6] Medical School, Tufts University, Boston, USA

Abstract. For over one hundred years the electrocardiogram (ECG) remains an extremely useful clinical tool and continues to play a major role in the evaluation and management of patients with known or suspected cardiac disease. Interpretation of the 12-lead ECG is a simplistic solution to the "inverse electromagnetic problem" for the electrical activity of the heart, which is to extract information about the instantaneous electrical state of the cardiac muscle from measurements of the body surface potentials that are generated from the electrical activity of the heart. Although adequate for patient management in most instances, there are conditions for which the sensitivity of the 12-lead ECG is suboptimal, as for example for the diagnosis of a posterior wall myocardial infarction. To enhance the diagnostic value of the ECG, one would need to address in depth and provide an actual solution to the inverse electromagnetic problem. Such a formal scientific solution of the inverse problem is a rather complicated matter and it requires, apart from measurements of the body surface potentials, a detailed knowledge of:

1. Chest geometry and structure, including information about the anatomy, tissue anisotropy and structural inhomogeneities of the thorax.
2. The histology of the heart, mainly the myocardial fiber length, density and orientation, since the electromagnetic field that the heart generates is the result of numerous elementary dipolar excitations corresponding to the excitations of the individual myocardial fibers.
3. Information regarding the three-dimensional positional variations of the heart over time. These changes occur throughout the cardiac cycle, as the heart

E.K. Theofilogiannakos et al.: *Attacking the Inverse Electromagnetic Problem of the Heart with Computationally Compatible Anatomical and Histological Knowledge*, Studies in Computational Intelligence (SCI) **65**, 109–129 (2007)
www.springerlink.com

contracts and twists during ventricular systole and relaxes and untwists during diastole. The deformation of the heart results in alterations in the myocardial fiber length, density and orientation, but also affects the properties of the volume conductor (i.e. the conductivities and resistivities of the thoracic tissues). Finally, during free breathing the heart is displaced within the chest cavity, predominantly in the craniocaudal direction, following the diaphragmatic up-down respiratory motion. These positional variations significantly affect the body surface potentials generated by the cardiac electrical activity.

Irrefutable evidence for all these factors comes from the significant changes in the ECG recordings that are observed between the donor and the recipient of a heart transplant.

In this review we examine, from a clinician's standpoint, the basic principles of solving the inverse electromagnetic problem of the heart for actual body geometries, and discuss the parameters that affect body surface potentials. We also present the types and depth of computational intelligence that are needed to calculate the contribution of several anatomic and functional cardiac and chest parameters, for a precise solution of the inverse electromagnetic problem of the heart.

6.1 Introduction

In 1893 Willem Einthoven (1924 Nobel Prize in Physiology and Medicine) introduced the term "electrocardiogram" (ECG) at a meeting of the Dutch Medical Association [1]. Ten years later, in 1903, he devised the first string galvanometer, known as the Einthoven galvanometer; with this instrument he was able to measure and record changes of the electrical potential that is generated by the functioning heart [2]. Einthoven described recordings from 3 limbs, to which 3 more unipolar limb leads were later added by Wilson. In 1942, Emmanuel Goldberger added the six precordial chest leads that, after an evolution of approximately forty years, gave rise to the current 12-lead ECG (Fig. 6.1).

The ECG is a highly cost-effective, noninvasive, portable, patient-friendly and risk-free diagnostic tool that has no contraindications in patients with intact skin. ECG is useful for the initial assessment, diagnosis and management of patients with known or suspected heart disease. In particular ECG is essential for the diagnosis of acute coronary syndromes, unstable angina and myocardial infraction (Fig. 6.2). In patients with myocardial infarctions the ECG morphology, namely the presence or absence of ST elevation, determines the optimal therapy. ECG is also the single most important non-invasive test for diagnosis of arrhythmias and conduction abnormalities, but has also great value for assessment of cardiac chamber hypertrophy and enlargement (Fig. 6.3), pericarditis and many non-cardiac diseases such as pulmonary embolism, lung diseases, drug toxicity, acid-base disturbances, subarachnoid hemorrhage etc [3]. For clinical interpretation or "reading" of the 12-lead ECG the physician compares the ECG of a patient with known electrocardiographic patterns that have been connected with specific diseases. For example a patient presenting with an ECG that shows ST segment elevation in the II,

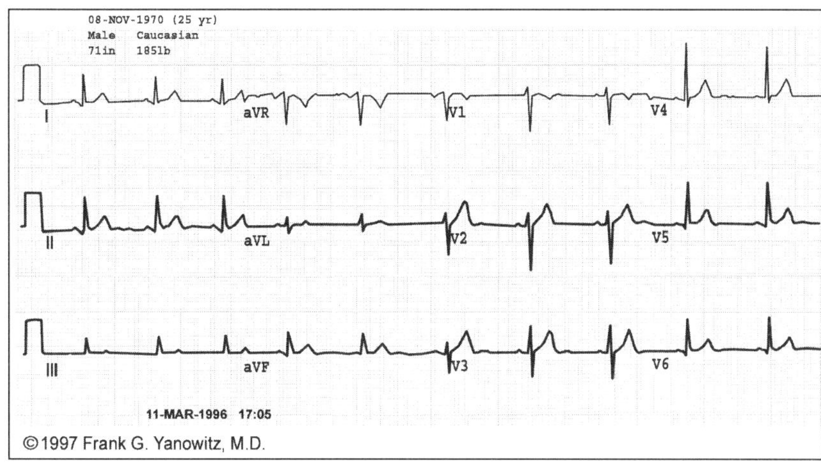

Fig. 6.1. Normal ECG trace as displayed by the "Alan E. Lindsay ECG learning center in Cyberspace". (http://library.med.utah.edu/kw/ecg/mml/ecg_12lead005.html last visited 29-8-2006). Permission to use granted 29-8-2006

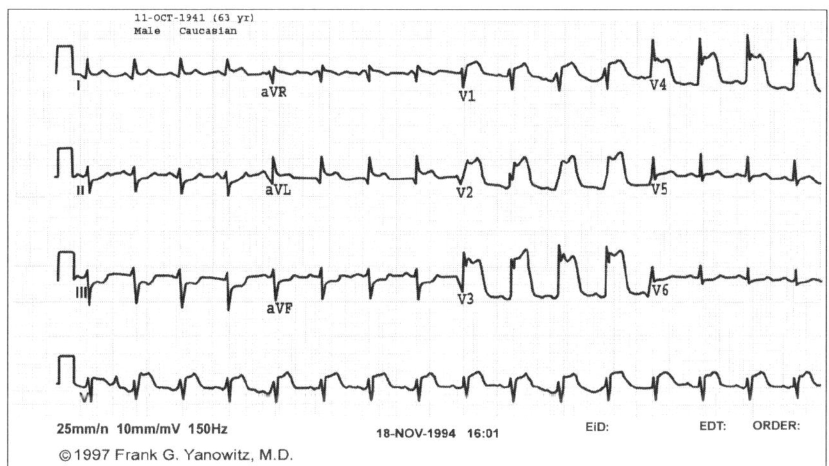

Fig. 6.2. Acute anterior myocardial infarction as displayed by the "Alan E. Lindsay ECG learning center in Cyberspace". (http://library.med.utah.edu/ kw/ecg/mml/ecg_12lead028.html last visited 29-8-2006). Permission to use granted 29-8-2006

III, aVF leads is diagnosed as suffering from a transmural inferior myocardial infarction.

However, since its early years, it has become evident that the value of the ECG is limited for certain conditions, for which both specificity and sensitivity may be suboptimal. For example, a prominent R wave in leads V1 and V2 may be seen both in patients with a posterior myocardial infraction, and also in patients with right ventricular hypertrophy, two entirely separate

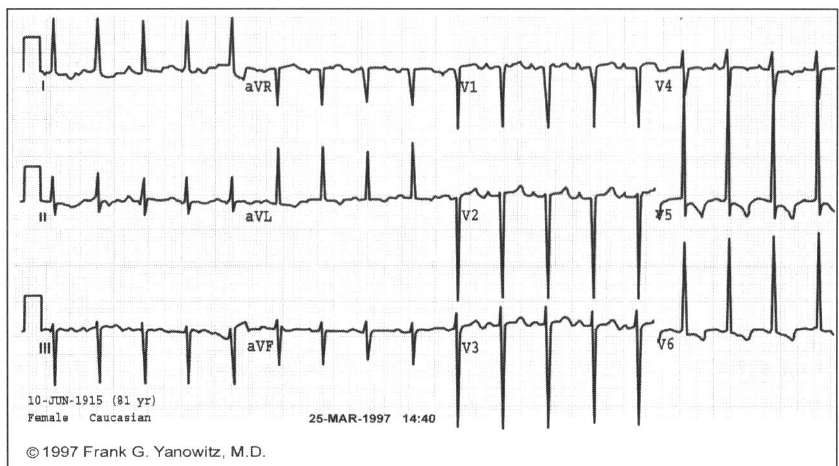

Fig. 6.3. Left Ventricular Hypertrophy with "Strain" as displayed by the "Alan E. Lindsay ECG learning center in Cyberspace". (http://library.med.utah.edu/ kw/ecg/mml/ecg_12lead024.html last visited 29-8-2006). Permission to use granted 29-8-2006

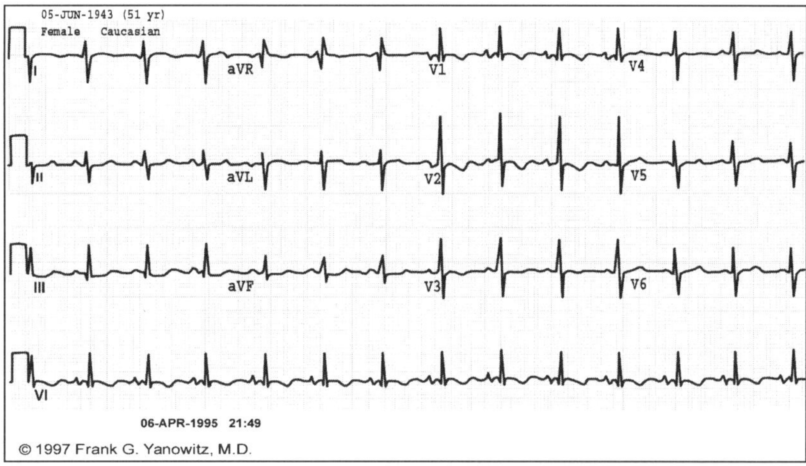

Fig. 6.4a. Severe Right Ventricular Hypertrophy as displayed by the "Alan E. Lindsay ECG learning center in Cyberspace" (http://library.med.utah.edu/kw/ ecg/mml/ecg_12lead064.html last visited 29-8-2006). Permission to use granted 29-8-2006

pathophysiologic entities (Figs. 6.4a and 6.4b) with different electrophysiological substrate, prognosis and treatment [4].

Another example where the ECG may not be diagnostic is the presence of complete or incomplete right bundle branch block with ST segment elevation in leads V1–V3, which on ECG grounds alone cannot distinguish between right ventricular myocardial infraction and Brugada syndrome. The electrophysiological

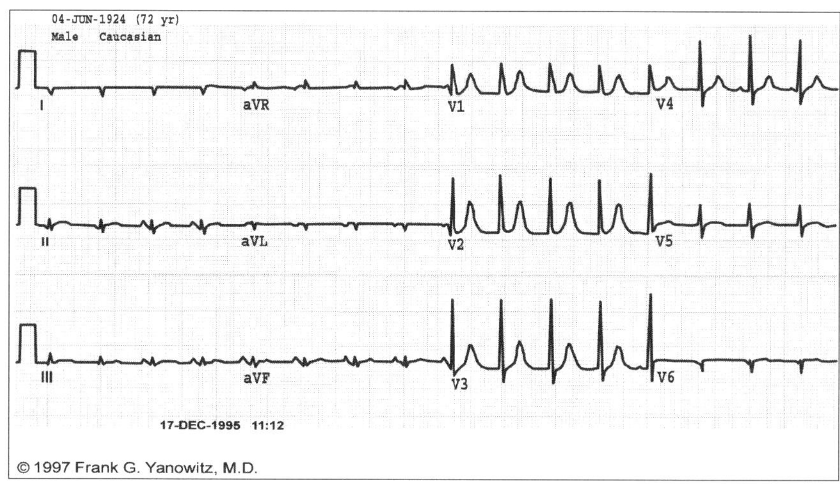

Fig. 6.4b. Fully Evolved Postero-lateral Myocardial Infarction as displayed by the "Alan E. Lindsay ECG learning center in Cyberspace". (http://library.med. utah.edu/kw/ecg/mml/ecg_12lead032.html last visited 29-8-2006). Permission to use granted 29-8-2006

substrate of the ST elevation in the case of the myocardial infraction is acute transmural myocardial ischemia/necrosis, in contrast to the sodium channel dysfunction that characterizes the Brugada syndrome [5], an autosomal dominant disease with incomplete penetrance that may cause syncope and sudden cardiac death in young individuals without structural heart disease [6]. Similarly, with left bundle branch block, the ECG has very low sensitivity and specificity for the diagnosis of ischemia and myocardial infarction, and also for the distinction between ischemic and nonischemic cardiomyoapthies.

Another limitation of the standard 12-lead ECG is that it represents a rather limited body-surface electrical mapping of cardiac electrical activity. It is thus inherently limited in identifying electrical signs from certain cardiac regions, not adequately "surveyed" by the conventional ECG leads, which remain "silent". One solution to this inadequate sampling would be to introduce new, optimally placed, additional leads (which are most sensitive to the electrophysiology of the "silent" heart regions) to monitor the cardiac electrical activity. To identify the optimal positioning of these new leads, one would have to know which areas on the chest wall have electrical activity that best describes the action potentials in the "silent" myocardial regions. Limitations of the standard 12-lead ECG could be overcome with the solution of the so-called inverse problem of electrocardiography.

6.2 The Inverse Problem

As inverse problem we define the process of inferring the distribution of an inaccessible variable from an accessible one [7], or, in other words identifying the unknown causes of known consequences. In ECG terms, the inverse problem is defined as the determination of the source (heart bound) from (a) the field that it impresses on the body surface and (b) the geometry of the volume conductor (thorax) through which the field spreads [8]. The first step in the direction of solving the inverse problem is to address and solve the forward problem. The forward problem is defined as the calculation of the electrical potential that is generated by a known source at a site (or sites) within, or at the surface of a volume conductor distant from the known source (Fig. 6.5). For electrocardiography, the forward problem would be to calculate the body surface potentials at various sites on the skin, for a given cardiac electrical activity. The solution of the forward problem does not have per se any application in clinical practice.

The two important applications of the inverse problem in electrocardiology are in the areas of ECG and in magnetocardiography (MCG). It is solved informally thousands of time per day, whenever a cardiologist makes an inference about the cardiovascular status of patients by reading their ECG. In fact "reading the ECG" is an empirical solution of inverse problem; the cardiologist compares the ECG of a patient with known electrocardiographic patterns that have been connected with specific diseases.

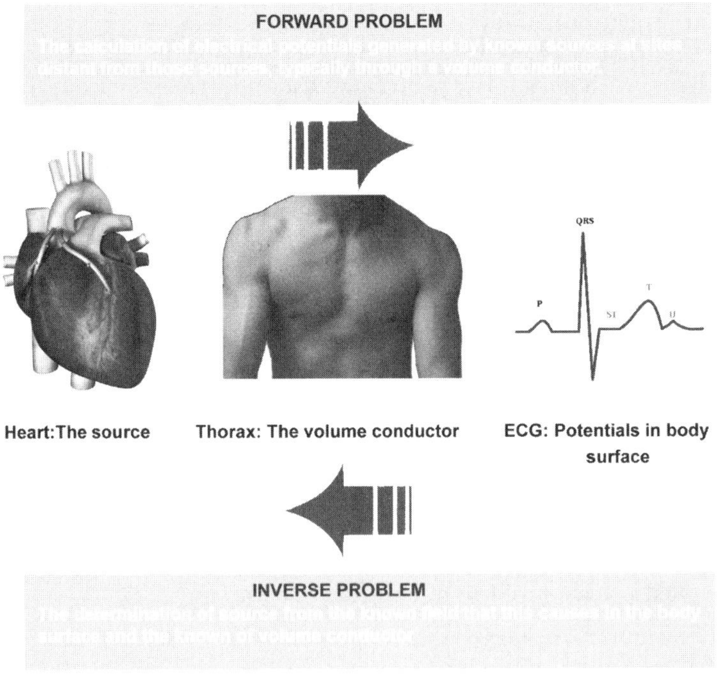

Fig. 6.5. The essence of the forward and the inverse electromagnetic problem for the heart

A major difference between the forward and inverse problems is that the former has a single solution, while the latter does not have a unique mathematical solution [9]. In this sense the inverse problem is an ill-posed problem, and small perturbations in the measured data (e.g., due to noise, errors in the forward model etc.) can result in unbounded errors in the inverse solution [10]. The objective of this chapter is not to review the numerous mathematical formulas that have been proposed for this purpose, but to examine the basic principles that must govern the solution of the inverse electromagnetic problem of the heart, considering the critical parameters (geometrical, histological, electrical, motional) that characterize both the source (heart) and volume conductor (thorax) of the cardiac electromagnetic field.

6.3 How Many Leads and Where?

Body Surface Potential versus Standard ECG

Until the mid 90s solutions of the inverse electromagnetic problem of the heart were based on the hypothesis that the heart, the source of the electromagnetic field, could be accurately described by a single electrical dipole located near the center of the heart [11]. The orientation and magnitude of the dipole was thought to vary during the cardiac cycle. If this assumption was a sufficient description of the source of the cardiac electric field, then just two nonparallel dipolar leads in the frontal plane would contain all the significant information about the electrical activity of the heart and would be sufficient to define the projection on the frontal plane of the dipole electrical moment (heart vector) of the heart. Furthermore, this model (which, in fact, is the model of the electrical activity of the heart that is implicit in Einthoven's formulation), implies that the use of a third lead out of the frontal plane enables the computation of the spatial magnitude and orientation of the heart vector [12]. Today, this single fixed-dipole model is no longer accepted as a sufficient description of the cardiac electrical activity. Instead, the multiple dipole model, which requires cardiac signal recording from many more leads than those used for the standard ECG, is in vogue (vide infra). Still, in clinical practice the 12-lead ECG is the gold standard and the cornerstone of diagnosis of cardiovascular disease. In the case of the 12-lead ECG, each lead always "faces", albeit from a different and distinct vantage point, the same electrophysiological event, i.e., the depolarization and repolarization of the heart muscle. However, the standard 12-lead ECG uses only nine such vantage points to record the electrical signals generated by the heart. Therefore, because of the limited number and particular placement of its leads, the standard 12-lead ECG fails to interrogate many areas of the myocardium [13]. As a result the sensitivity of 12-lead ECG is poor when it comes to the diagnosis of posterior wall or right ventricular myocardial infraction, because the right ventricular and posterior walls are electrically "silent" to the standard ECG [14].

This handicap is overcome in part by using the right sided (V4R, V5R, V6R) [15] and posterior leads (V7, V8, V9) [16]. These extra leads "view" the right ventricle and the posterior left ventricular wall, offering direct information for the electrical status of these areas. In fact many studies support the replacement of the standard 12-lead ECG by an 18-lead ECG in clinical practice [17].

To further increase the detail and accuracy of ECG, recording with a large number of leads placed over the entire chest wall (Body Surface Potential Mapping system or BSPMs – Fig. 6.6) has been proposed. Systems with 32, 64, 128 or 256 leads have been described [18]. Although it failed to gain broad acceptance, BSPM has been the first method ever used in human electrocardiography [19].

The main advantage of BSPM is that it allows high spatial resolution of cardiac electrical events [21]. The clinical utility of BSPM could include:

(1) The accurate detection of acute myocardial infarction. Table 6.1 includes a list of studies that demonstrate superiority of BSPM compared

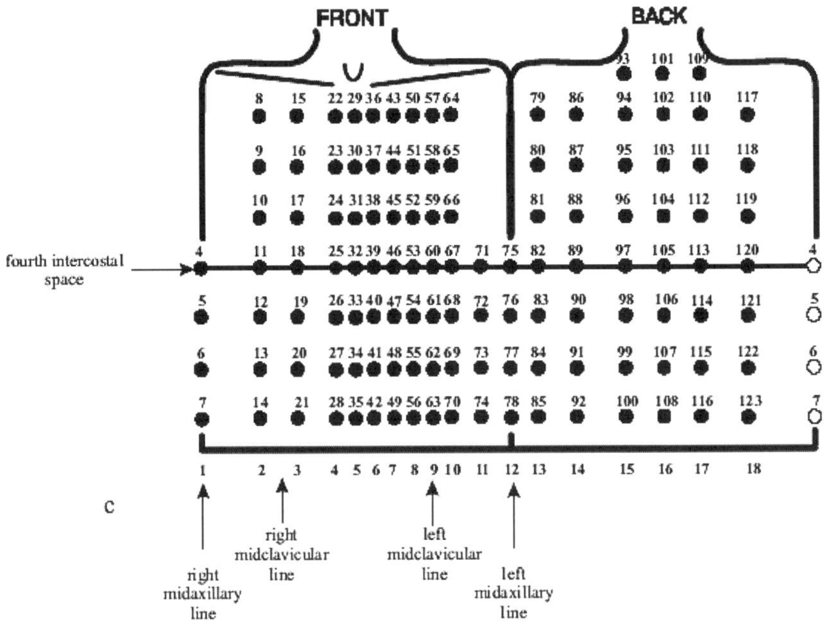

Fig. 6.6. Body Surface Potential Mapping electrode layout. In this case, the 120 electrodes are mounted on 18 vertical strips. According to the author, Helena Hänninen [20], the dimensions of the upper body determine horizontal spacing of the electrodes. Strip 1 is located on the right and strip 12 on the left mid-axillary line, respectively. Strip 9 is located on the left mid-clavicular line. The right mid-clavicular line is between strips 2 and 3. In all strips, the fourth electrode from the inferior end of the strip was placed at the level of the fourth inter-costal space determined at the border of the sternum. (http://ethesis.helsinki.fi/julkaisut/laa/kliin/vk/hanninen/ last visited 29-8-2006)

Table 6.1. List of papers that advocate the superiority of BSPM to standard 12-lead ECG for diagnosis of acute myocardial infarction (MI)

Author	Year of publication	Symptoms-Disease	Number of Patients	Number of BSPM leads	12-lead ECG		BSPM	
					Specificity	Sensitivity	Specificity	Sensitivity
Carley et al [28]	2005	MI symptoms	389		94%	40%	85%	47%
Owens et al [29]	2004	Angina symptoms	294	80	94%	57%	92%	80%
McClelland et al [30]	2003	Angina symptoms without ST changes	103	80	98%	45%	94%	64%
Maynard et al [23]	2003	Angina symptoms and LBBB	56	80	87%	17%	71%	67%
Menown et al [31]	2001	Angina and ST depression	54	80	81%	38%	75%	88%
Green et al [32]	1987	Coronary disease	40				98%	94%
Kornreich et al [33]	1986	MI anterior-lateral	114	120		89%	95%	97%
Kornreich et al [33]	1986	MI posterior	114	120		85%	95%	94%

Table 6.2. Disadvantages of BSPM systems

---

1 The cost of BSPM is substantially higher than that of the 12-lead ECG.

2 There is lack of standardization of lead positions and/or necessity to identify the exact coordinates of each lead, which makes the system cumbersome and unfit for routine clinical work.

3 BSPM demands substantially more time for its interpretation while a method for its automatic and unequivocal interpretation has yet to be devised.

4 The specificity of the BSPM is inferior to that of the 12-lead ECG, resulting frequently in over-diagnosis.

---

to standard 12-lead ECG for diagnosis of acute myocardial infarction, particularly in patients with chest pain without diagnostic ECG changes [22] or in patients with chest pain and left bundle branch block [23].

(2) The identification and precise localization of bypass tracts in patients with ventricular pre-excitation and identifiable delta waves [24].

(3) The detection and localization of late potentials which have been connected with an increased risk of ventricular arrhythmias [25].

(4) The detection of localized conduction disturbances [26].

(5) The detection of ventricular hypertrophy [27].

During the last three decades many studies support that BSPM has higher diagnostic accuracy than the standard 12-lead ECG, offering more detailed regional information for cardiac electrical activity. Despite its purported diagnostic advantages, BSPM has not replaced or supplanted the ECG due to its several disadvantages (Table 6.2). To this date, BSPM remains largely an experimental tool without any significant clinical use [34]. It is likely that further development of BSPM will depend on parallel developments in the solution of the inverse electromagnetic problem of the heart, incorporating detailed knowledge of the source properties (i.e., of the heart anatomy, histology, electrical properties and changes in shape and position of the heart muscle during the cardiac cycle) as well as a detailed knowledge of the volume conductor where the heart is embedded (i.e., of the geometry, motion and tissue conductivities of the thorax).

6.4 Heart- The Source of the Field

Aristotle believed that the heart is the seat of the soul. Two and a half thousand years later, heart electrophysiology and electrocardiology are major clinical and research areas of modern cardiology and bioengineering. Excitable tissues such as heart muscle mediate both the communication among their cells and the coordination of their contractile activity through electrical currents. These endogenous currents are capable of generating electromagnetic fields sufficiently large to be measured outside of the body [35]. The cardiac electromagnetic field has two components: the electric field, which is recorded by ECG, and the magnetic field that can be recorded by the magnetocardiogram (MCG) [35,36]. These two components are generated by the same event, namely the electric activity of the heart, and thus are not separate. Based on the physics of the electromagnetic field, four basic models have been proposed to describe the heart as the volume source of the electromagnetic field: (1) the Single Fixed Dipole model, (2) the Moving Dipole model, (3) the Multiple Dipole model and (4) the Multipole model.

In the case of the Single Fixed Dipole model, which dates back to at least the 1950's [37], the heart, as a source, is assumed to be characterized by a current dipole M, which is located at the center of the heart's volume. This dipole has fixed location but its magnitude and direction varies. Hence the description of heart sources is degenerated in the determination of three unknowns, namely, the Cartesian moments M_x, M_y, M_z of the dipole M. Each of these unknown and independent variables is the magnitude of the dipole at x, y and z direction respectively, which describe in a unique way the fixed current dipole.

The Moving Dipole model is the natural first order generalization of the Single Fixed Dipole model in the sense that it can be generated from it by varying not only the magnitude and direction but also the location of its origin [38]. Therefore, except for the Cartesian moments M_x, M_y, M_z the new location ($x_{new}, y_{new}, z_{new}$) of the dipole M has to be determined. That means that the original three independent variables that characterize the Single Fixed Dipole model, grow to six in order to account for the motion of the dipole's origin.

In a similar manner, the Multiple Dipole model [39] includes several dipoles, each fixed in its location and representing a certain anatomical region of the heart. Hence the locations of those dipoles are fixed and only their magnitude and direction varies. In the case of the Multiple Dipole model, the number of the variables becomes equal to $3xn$, where n is the total number of the dipoles used to describe the heart. If, in addition to the origin, the assumption of fixed direction is also applied, then it becomes clear that only one independent variable, the magnitude of each dipole, has to be estimated. Therefore the number of the independent variables for multiple dipoles with fixed orientation can be reduced to that of the number of the dipoles n.

Finally, according to the Multipole model a dipole can be formed by two equal and opposite monopoles placed close together, a quadrupole can be formed by two equal and opposite dipoles placed close together, an octapole can be formed by the placement of two quadrupoles close together, and so on. The advantage of this model is that any given source configuration can be expressed as an infinite sum of multipoles of increasing order (i.e., dipole, quadrupole, octapole, etc.). Thus, in principle, by choosing ever higher order multipoles, one can arrive to as accurate a description of the source as one desires [40]. The size of each component multipole depends on the particular source distribution. For the full definition of a dipole three variables have to be determined. For the definition of a quadrupole, which is the combination of two dipoles, five variables are necessary: three describe the first dipole and the other two are used to define the location of the second dipole, making the assumption that the magnitude of the two component dipoles of the quadrupole is equal. In the same way, the octapole presents as the combination of two quadrupoles and it can be defined by seven coefficients. Five of them describe the first quadrupole while the other two are necessary to define the location of the second quadrupole. It is evident, therefore, that geometrical, anatomical and histological parameters of heart are important factors that determine the electrical field that it generates. Therefore, such parameters are extremely important, not only in a physical/physiological sense, but also in a biological sense, for the solution of the forward and inverse problems.

The importance of the geometrical, anatomical and histological cardiac and chest parameters in solving the inverse electromagnetic problem is underlined by the observation that differences in the heart position and orientation within the thorax result in amplitude differences of the QRS complex of the 12-lead ECG and these differences constitute part of the observed interindividual variability of the ECG [41]. Furthermore, as the heart contracts during systole, it also rotates and produces torsion due to the structure of the myocardium [42]. This, in turn, gives rise to a shift of the cardiac dipole source location compounded by the fact that, when the heart volume changes, the extent to which such a change is accompanied by a positional change strongly influences the surface voltage. In fact, the surface voltage may increase or decrease following an increase in heart volume [43]. Three-dimensional cardiac geometry and motion that affect the heart shape and its relative position and orientation to the whole body, can be imaged by novel imaging techniques [44], such as cardiac MRI or CT.

The electromagnetic field that the heart generates is the result of numerous elementary dipolar excitations corresponding to the excitation of the individual myocardial fibers. Therefore, the knowledge of cardiac microstructure (fiber density and orientation) is necessary for an accurate description of the field source and a prerequisite for any large scale simulation of the electrical and mechanical behavior of the heart. The cardiac anisotropy, which is directly related to its microstructure, affects the generation of BSPM in two respects: firstly the anisotropic conduction of the cardiac action potential and secondly the anisotropic distribution of the electrical conductivity of the heart tissue [45]. This has created the hypothesis that cardiac anisotropy significantly affects the propagation of excitation wave fronts within the ventricles irrespective of the fact that its effect on the inverse solution from BSPM is tolerable [46]. Although the validity of this hypothesis has not been verified, the effect of myocardial fiber direction on epicardial potentials has been demonstrated in a canine model [47].

Diffusion Tensor MRI (DTMRI) is a novel imaging technique that offers high resolution images of cardiac fiber orientation [48]. In DTMRI the assignment of the local fiber direction is based on the assumption that the diffusion of water molecules by Brownian motion is larger in the longitudinal direction of the cells than in a direction that is transverse to it. In general the Brownian motion of water molecules is greater along fibers than across fibers, leading to an anisotropic diffusion tensor [49]. However, cardiac microstructure is not static. During the continuous heart motion (systole-diastole cycle) the density, length and orientation of its fibers continuously change. The strain of muscle fibers in the heart is likely to be distributed uniformly over the cardiac walls during the ejection period of the cardiac cycle [50]. The spatial distributions of fiber stress and strain in the cardiac walls have also been predicted with sufficient accuracy by mathematical models. Knowledge of these models, coupled with detailed DTMRI images, may further refine the quantification of the parameters necessary for solving the inverse electromagnetic problem of the heart and may result in further refinement of the solution through the simulation of the changes that occur in fiber orientation during the systole-diastole cycle.

6.5 Thorax- The Volume Conductor of the Field

Because the heart is located within the torso (which forms the intermediate medium between the cardiac source and the body surface), body surface potentials are influenced by torso geometry and position as well as by torso inhomogeneities. One would thus expect that the same heart would provide different ECG tracings in different chest cavities, as is indeed the case with cardiac transplant donors and recipients. Cardiac transplantation provides a unique opportunity to record the electric field generated by a human heart within a different volume conductor and it has been proposed that the differences between pre – and post – transplantation ECGs are largely due to their differences in thorax geometry and anatomy [51]. Several studies have assessed the influence of tissue inhomogeneities and anisotropies on the electric field or the magnetic field of the heart [52]. Some investigators suggest that effects of torso inhomogeneities on body surface potential are minimal [53] while others claim that the accuracy of tissue inhomogeneity representation has a significant effect on the accuracy of the forward and inverse solution. In general, regions near the torso surface seem to play a larger role than regions near the heart. Therefore, subcutaneous fat, skeletal muscle and the size of the lungs apparently have a larger impact on the body – surface potential distribution than the intracardiac blood pool and the epicardial fat [54, 55]. Recently, it was shown that the influence of different tissue inhomogeneities on the body – surface potentials that are generated by atrial electrical activity is significant and it was suggested that inhomogeneities, especially those involving the lungs and the atrial cavities, must be incorporated in the different volume conductor models that link atrial electric activity to body surface potentials [56]. The influence of the various thoracic tissue conductivities on body surface potential is also suggested from certain ECG findings in particular pathological conditions. In chronic obstructive pulmonary disease, for example, which is characterized by decreased lung electrical conductivity due to over inflation of the lung, the ECG amplitude is augmented [57]. In contrast, in the case of acute pulmonary edema, which is characterized by increased lung electrical conductivity due to alveolar liquid accumulation, the surface ECG amplitude is decreased.

The accuracy of both the forward and the inverse problem solutions relies on the precise modeling of the thorax and good knowledge of its electrical anisotropy. The heart is surrounded by the torso which is made up by many different tissues, each with a different and characteristic conductivity (Figs. 6.7a and 6.7b) that create the electrical resistivity inhomogeneity (Resistivity = 1/conductivity). Table 6.3 lists the resistivity values of thoracic tissues as previously reported, but which vary significantly among studies [58]. A possible explanation for the apparently wide variation in the reported values is that the resistivity of the blood largely depends on the haematocrit, the pH and blood urea [59]. As the resistivity of blood is directly proportional to the haematocrit, an increase in the haematocrit is expected to result in a decrease in the ECG amplitude. Thus, in patients with anemia, the ECG amplitude is expected to increase, while in polycythemia (increased number of red blood cells in a given volume of blood), a decrease in the ECG amplitude

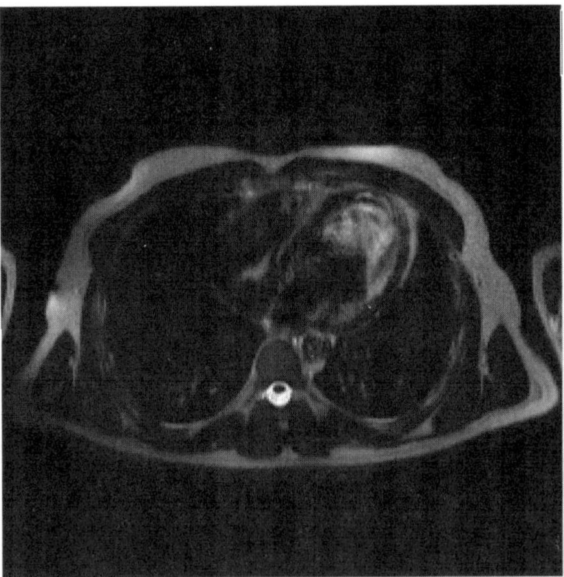

Fig. 6.7a. The accuracy of both the forward and the inverse problem solutions is based on the accuracy of the thorax model and identification of its regions of uniform resistivity. The MRI is a vital tool for visualizing the anatomy of the torso and delineating the borders of the different tissues that make up the torso

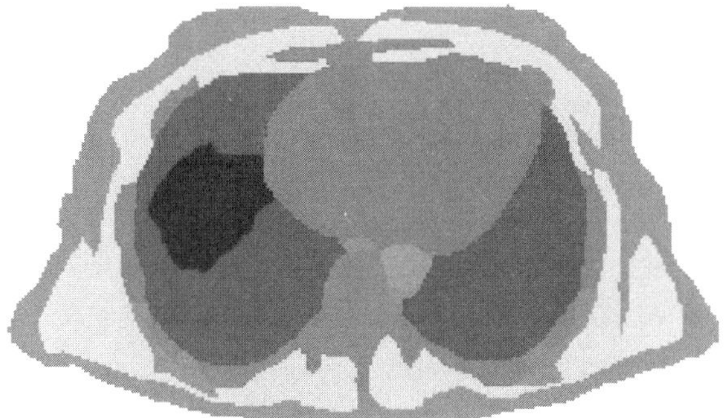

Fig. 6.7b. Once the anatomy of the torso structures (and/or of the tissues that make them up) has been delineated, it is possible to identify regions of uniform resistivity within each section of the torso. The identification is based on the assumption that solid organs are made of tissue with uniform resistivity. In the example on the left, the heart is considered of uniform resistivity mainly because its chambers are always filled with blood and due to the fact that the resistivity of the heart muscle is of the same order of magnitude as the resistivity of blood (Table 6.3)

Table 6.3. Chest tissue resistivity

Tissue	Resistivity (Ω)
Blood	1.6
Heart muscle	2.5 (parallel)
	5.6 (normal)
Skeletal muscle	1.9 (parallel)
	13.2 (normal)
Lungs	20
Fat	25
Bone	177

is expected [60] The conductivity of other tissues is strongly influenced by many physiological changes, such as postural changes [61], or by pathological conditions, such as disturbances in the electrolyte or water balances of the body. For example the increase in the QRS amplitude after hemodialysis is influenced by the changes of the conductivity of extracardiac thorax and the relative position of the heart within the chest cavity [62].

Besides the need for information regarding the material properties and inhomogeneities of the thoracic tissues, the solution of both the forward and the inverse problem requires a detailed image of thoracic cavity. The required high spatial and contrast resolution can be obtained from noninvasive imaging techniques such as CT and MRI (Figs. 6.8a and 6.8b). The three dimensional reconstructions obtained from contiguous two dimensional images have proved

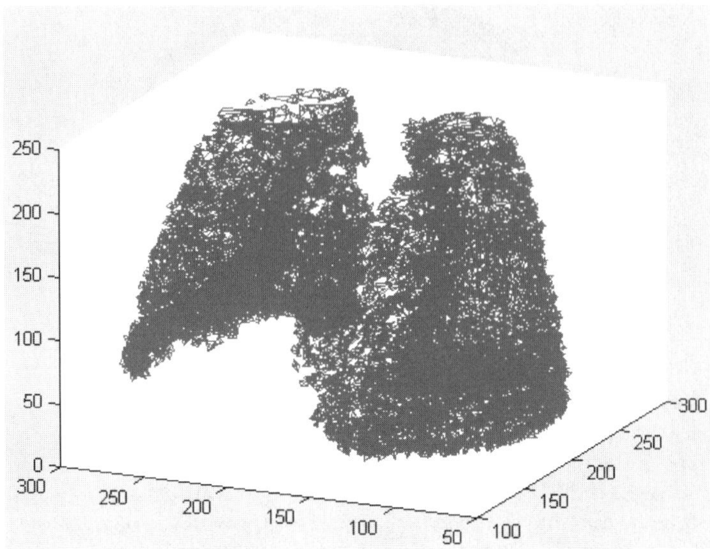

Fig. 6.8a. 3D reconstruction, from 2D MRI images (same patient as the one displayed in Fig. 6.6a). In this picture (which is displayed on the set of software tools that the authors routinely use) the outline of the lung can be discerned with the heart removed from their midst. Such reconstructions will greatly facilitate the solution of both the forward and the inverse problem in electrocardiography

to be an effective way to assess the anatomic relationship between the different structures within the thorax and their relative displacement during the cardiac cycle [63].

6.6 Conclusions

Accurate solution of the inverse problem of electrocardiography would significantly increase the diagnostic accuracy of ECG. Solution of this problem would require sampling of the electrical activity of the heart from multiple locations on the chest wall, the optimal selection of which can be determined from solution of the forward problem of electrocardiography. In addition, knowledge of cardiac histology, anatomy and physiology as well as information regarding chest anatomy and tissue resistivity are essential variables to be considered. This information can currently be derived from imaging technology. Progress in computer technology has increased our ability to undertake the solution of immensely complex mathematical problems and to consider the contribution of multiple variables for the solution of the inverse problem. The next frontier, in this direction, is the integration of detailed anatomical and histological information regarding the heart and chest, into the solution of the inverse electromagnetic problem of the heart. Incorporation of cardiac

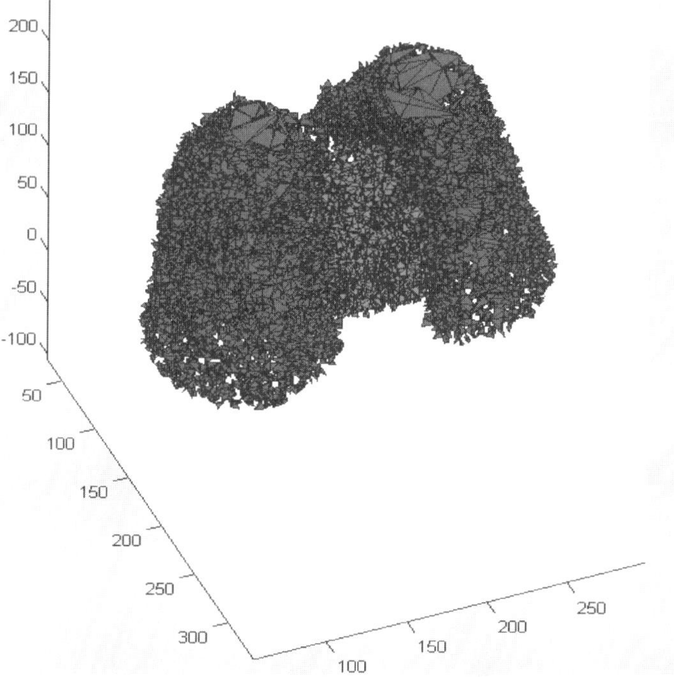

Fig. 6.8b. Another view of the same structures as in Fig. 6.8a. This time the heart (red) volume is included in the rendering

motion would be an even more distant step in this direction. These advances could provide an individualized and highly accurate way of assessing cardiac physiology and pathology, and possibly a novel non-invasive clinical tool.

References

1. Einthoven, W.: Nieuwe methoden voor clinisch onderzoek [New methods for clinical investigation]. Ned. T. Geneesk. 29 II (1893) 263–286
2. Einthoven, W.: Galvanometrische registratie van het menschilijk electrocardiogram. In: Rosenstein, S.S. (ed.): Herinneringsbundel. Eduard Ijdo, Leiden (1902)101–107
3. Kadish, A.H., Buxton, A.E., Kennedy, H.L., Knight, B.P., Mason, J.W., Schuger, C.D., Tracy, C.M., Boone, A.W., Elnicki, M., Hirshfeld, J.W., Lorell, B.H., Rodgers, G.P., Tracy, C.M., Weitz, H.H.: ACC/AHA Clinical Competence Statement on Electrocardiography and Ambulatory Electrocardiography. J.A.C.C. 38 (2001) 2091–2100
4. Mattu, A., Brady, W.J., Perron, A.D., Robinson, D.A.: Prominent R wave in lead V1: electrocardiographic differential diagnosis. Am. J. Emerg. Med. 19 (2001) 504–513

5. Brugada, P., Brugada, R., Antzelevitch, C., Brugada, J.: The Brugada Syndrome. Arch. Mal. Coeur. Vaiss. 98 (2005) 115–122
6. Francis, J., Antzelevitch, C.: Brugada syndrome. Int. J. Cardiol. 101 (2005) 173–178
7. Smith, W.M., Barr, R.C.: The Forward and Inverse problems: What are they, why are they important, and where do we stand? J. Cardiov. Electrophysiol. 12 (2001) 253–255
8. Malmivuo, J., Plonsey, R.: Bioelectromagnetism Principles and Applications of Bioelectric and Biomagnetic Fields. Oxford University Press, New York (1995).
9. Gulrajani, R.M.: The forward and inverse problems of electrocardiography. IEEE Eng. Med. Biol. Mag.17 (1998) 84–101
10. Ahmad, G.F., Brooks, D.H., MacLeod, R.S.: An admissible solution approach to inverse electrocardiography. Ann. Biomed. Eng. 26 (1998) 278–292
11. Lepeschkin, E.: History of electrocardiographic thoracic mapping. In Rush, S., Lepeschkin, E. (eds): Body Surface Mapping of Cardiac Fields. Karger, Basel, (1974) 2–10
12. Savjaloff, V.: Methode der stereometrischen Elektrokardiographie. Z. Kreislaufforsch 21 (1929) 705
13. Carley, S.D.: Beyond the 12 lead: review of the use of additional leads for the early electrocardiographic diagnosis of acute myocardial infarction. Emerg. Med. (Fremantle). 15 (2003) 143–154
14. Perloff, J.K.: The recognition of strictly posterior myocardial infarction by conventional scalar electrocardiography. Circulation 30 (1964) 706–718
15. Braat, S.H., Brugada, P., den Dulk, K., van Ommen, V., Wellens, H.J.: Value of lead V4R for recognition of the infarct coronary artery in acute inferior myocardial infarction. Am. J. Cardiol. 53 (1984) 1538–1541
16. Matetzky, S., Freimark, D., Feinberg, M.S., Novikov, I., Rath, S., Rabinowitz, B., Kaplinsky, E., Hod, H.: Acute myocardial infarction with isolated ST-segment elevation in posterior chest leads V7-9: "hidden" ST-segment elevations revealing acute posterior infarction. J. Am. Coll. Cardiol. 34 (1999) 748–753
17. Stewart, S., Haste, M.: Prediction of right ventricular and posterior wall ST elevation by coronary care nurses: the 12-lead electrocardiograph versus the 18-lead electrocardiograph. Heart-Lung 25 (1996) 14–23
18. Green, L.S., Abildskov, J.A.: Clinical applications of body surface potential mapping. Clin. Cardiol. 18 (1995) 245–249
19. Waller, A.: On the electromotive changes connected the beat of mammalian heart and of the human heart in particular. Philos. Trans. R. Soc. B (1889) 180–189
20. Hänninen, H.: Multichannel magnetocardiography and body surface potential mapping in exercise-induced myocardial ischemia. Academic Dissertation, Medical Faculty of the University of Helsinki, Helsinki (2002)
21. Nadeau, R., Savard, P., Gulrajani, R., Cardinal, R., Molin, F., Page, P.: Clinical applications of BSM. J. Electrocardiol. 28 (1995) 334–335
22. Menown, I.B., Patterson, R.S., MacKenzie, G., Adgey, A.A.: Body-surface map models for early diagnosis of acute myocardial infarction. J. Electrocardiol. 31 Suppl (1998) 180–188
23. Maynard, S.J., Menown, I.B., Manoharan, G., Allen, J., McC Anderson, J., Adgey, A.A.: Body surface mapping improves early diagnosis of acute myocar-

dial infarction in patients with chest pain and left bundle branch block. Heart 89 (2003) 998–1002

24. Giorgi, C., Nadeau, R., Savard, P., Shenasa, M., Page, P.L., Cardinal, R.: Body surface isopotential mapping of the entire QRST complex in the Wolff-Parkinson-White syndrome. Correlation with the location of the accessory pathway. Am. Heart. J. 121 (1991) 1445–53

25. Shibata, T., Kubota, I., Ikeda, K., Tsuiki, K., Yasui, S.: Body surface mapping of high-frequency components in the terminal portion during QRS complex for the prediction of ventricular tachycardia in patients with previous myocardial infarction. Circulation 82 (1990) 2084–2092

26. Mirvis, D.M.: Conduction abnormalities and ventricular hypertrophy. In: Mirvis, D.M. (eds): Body surface electrocardiographic mapping. Kluwer Academic Publishers, Boston (1988) 153

27. Kornreich, F., Montague, T.J., Rautaharju, P.M., Kavadias, M., Horacek, M.B.: Identification of best electrocardiographic leads for diagnosing left ventricular hypertrophy by statistical analysis of body surface potential maps. Am. J. Cardiol. 62 (1988) 1285–1291

28. Carley, S.D., Jenkinsm, M., Jones, K.M.: Body surface mapping versus the standard 12 lead ECG in the detection of myocardial infarction amongst Emergency Department patients: a Bayesian approach. Resuscitation 64 (2005) 309–314

29. Owens, C.G., McClelland, A.J., Walsh, S.J., Smith, B.A., Tomlin, A., Riddell, J.W., Stevenson, M., Adgey, A.A.: Prehospital 80-LAD mapping: does it add significantly to the diagnosis of acute coronary syndromes? J. Electrocardiol. 37 Suppl (2004) 223–232

30. McClelland, A.J., Owens, C.G., Menown, I.B., Lown, M., Adgey, A.A.: Comparison of the 80-lead body surface map to physician and to 12-lead electrocardiogram in detection of acute myocardial infarction. Am. J. Cardiol. 92 (2003) 252–257

31. Menown, I.B., Allen, J., Anderson, J.M., Adgey, A.A.: ST depression only on the initial 12-lead ECG: early diagnosis of acute myocardial infarction. Eur. Heart J. 22 (2001) 218–227

32. Green, L.S., Lux, R.L., Haws, C.W.: Detection and localization of coronary artery disease with body surface mapping in patients with normal electrocardiograms. Circulation 76 (1987) 1290–1297

33. Kornreich, F., Montague, T.J., Rautaharju, P.M., Block, P., Warren, J.W., Horacek, M.B.: Identification of best electrocardiographic leads for diagnosing anterior and inferior myocardial infarction by statistical analysis of body surface potential maps. Am. J. Cardiol. 158 (1986) 863–871

34. Taccardi, B., Punske, B.B., Lux, R.L., MacLeod, R.S., Ershler, P.R., Dustman, T.J., Vyhmeister, Y.: Useful lessons from body surface mapping. J. Cardiovasc. Electrophysiol. 9 (1998) 773–786

35. Plonsey, R.: Bioelectric Phenomena. McGraw-Hill, New York (1969)

36. Tavarozzi, I., Comani, S., Del Gratta, C., Romani, G.L., Di Luzio, S., Brisinda, D., Gallina, S., Zimarino, M., Fenici, R., De Caterina, R.: Magnetocardiography: current status and perspectives. Part I: Physical principles and instrumentation. Ital. Heart J. 3 (2002) 75–85

37. Frank, E.: An accurate, clinically practical system for spatial vectocardiography. Circulation 13 (1956) 737

38. Gulrajani, R.M., Roberge, F.A., Savard, P.: Moving dipole inverse ECG and EEG solutions. IEEE Trans. Biomed. Eng. 31 (1984) 903–910

39. Miller, W.T. 3rd., Geselowitz, D.B.: Use of electric and magnetic data to obtain a multiple dipole inverse cardiac generator: a spherical model study. Ann. Biomed. Eng. 2 (1974) 343–360
40. Beetner, D.G., Arthur, R.M.: Estimation of heart-surface potentials using regularized multipole sources. IEEE Trans. Biomed. Eng. 51 (2004) 1366–1373
41. Huiskamp, G.J., van Oosterom, A.: Heart position and orientation in forward and inverse electrocardiography. Med. Biol. Eng. Comput. 30 (1992) 613–620
42. Burleson, K.O., Schwartz, G.E.: Cardiac torsion and electromagnetic fields: the cardiac bioinformation hypothesis. Med. Hypotheses. 64 (2005) 1109–1116
43. Amoore, J.N.: The Body effect and change of volume of the heart. J. Electrocardiol. 18 (1985) 71–75
44. Petitjean, C., Rougon, N., Cluzel, P.: Assessment of myocardial function: a review of quantification methods and results using tagged MRI. J. Cardiovasc. Magn. Reson. 7 (2005) 501–516
45. He, B., Li, G., Zhang, X.: Noninvasive imaging of cardiac transmembrane potentials within three-dimensional myocardium by means of a realistic geometry anisotropic heart model. IEEE Trans. Biomed. Eng. 50 (2003) 1190–1202
46. Liu, C., He, B.: Effects of cardiac anisotropy on three-dimensional ECG localization inverse solutions: a model study. I.J.B.E.M. 7 (2005) 221–222
47. Taccardi, B., Macchi, E., Lux, R.L., Ershler, P.R., Spaggiari, S., Baruffi, S., Vyhmeister, Y.: Effect of myocardial fiber direction on epicardial potentials. Circulation 90 (1994) 3076–3090
48. Helm, P., Beg, M.F., Miller, M.I., Winslow, R.L.: Measuring and mapping cardiac fiber and laminar architecture using diffusion tensor MR imaging. Ann. N. Y. Acad. Sci. 1047 (2005) 296–307
49. Holmes, A.A., Scollan, D.F., Winslow, R.L.: Direct histological validation of diffusion tensor MRI in formaldehyde-fixed myocardium. Magn. Reson. Med. 44 (2000) 157–161
50. Rijcken, J., Bovendeerd, P.H., Schoofs, A.J., van Campen, D.H.: Arts T. Optimization of cardiac fiber orientation for homogeneous fiber strain during ejection. Ann. Biomed. Eng. 27 (1999) 289–297
51. Butman, S.M., Phibbs, B., Wild, J., Copeland, J.G.: One heart, two bodies: insight from the transplanted heart and its new electrocardiogram. Am. J. Cardiol. 66 (1990) 632–635
52. Bruder, H., Scholz, B., Abraham-Fuchs, K.: The influence of inhomogeneous volume conductor models on the ECG and the MCG. Phys. Med. Biol. 39 (1994) 1949–1968
53. Ramanathan, C., Rudy, Y.: Electrocardiographic imaging: I. Effect of torso inhomogeneities on body surface electrocardiographic potentials. J. Cardiovasc. Electrophysiol. 12 (2001) 229–240
54. Klepfer, R.N., Johnson, C.R., Macleod, R.S.: The effects of inhomogeneities and anisotropies on electrocardiographic fields: a 3-D finite-element study. IEEE Trans. Biomed. Eng. 44 (1997) 706–719
55. Bradley, C.P., Pullan, A.J., Hunter, P.J.: Effects of material properties and geometry on electrocardiographic forward simulations. Ann. Biomed. Eng. 28 (2000) 721–741
56. Van Dam, P.M., van Oosterom, A.: Volume conductor effects involved in the genesis of the P wave. Europace 7 Suppl 2 (2005) 30–38

57. Ikeda, K., Kubota, I., Yasui, S.: Effects of lung volume on body surface electro-cardiogram. Isointegral analysis of body surface maps in patients with chronic pulmonary emphysema. Jpn. Circ. J. 49 (1985) 284–291
58. Faes, T.J., van der Meij, H.A., de Munck, J.C., Heethaar, R.M.: The electric resistivity of human tissues (100 Hz-10 MHz): a meta-analysis of review studies. Physiol. Meas. 20 (1999) R1-10
59. Fuller, H.D.: The contribution of blood chemistry to the electrical resistance of blood: an in vitro model. Acta. Physiol. Hung. 92 (2005):147–151
60. Oreto, G., Luzza, F., Donato, A., Satullo, G., Calabro, M.P., Consolo, A., Arrigo, F.: Electrocardiographic changes associated with haematocrit variations. Eur. Heart J. 3 (1992) 634–637
61. Hyttinen, J., Puurtinen, H.G., Kauppinen, P., Nousiainen, J., Laarne, P., Malmivuo, J.: On the effects of model errors on forward and inverse ECG problems I.J.B.E.M. 2 (2000) 13
62. Kinoshita, O., Kimura, G., Kamakura, S., Haze, K., Kuramochi, M., Shimomura, K., Omae, T.: Effects of hemodialysis on body surface maps in patients with chronic renal failure. Nephron 64 (1993) 580–586
63. Miquel, M.E., Hill, D.L., Baker, E.J., Qureshi, S.A., Simon, R.D., Keevil, S.F., Razavi, R.S.: Three- and four-dimensional reconstruction of intra-cardiac anatomy from two-dimensional magnetic resonance images. Int. J. Cardiovasc. Imaging 19 (2003) 239–254

Human Machine Interface for Healthcare and Rehabilitation

Giuseppe Andreoni[1], Sergio Parini[1], Luca Maggi[1], Luca Piccini[1],
Guido Panfili[1], and Alessandro Torricelli[1]

[1] Politecnico di Milano, P.zza L. Da Vinci 32, 20133 Milan, Italy
giuseppe.andreoni, luca.piccini, luca.maggi,
sergio.parini, alessandro.torricelli@polimi.it http://www.polimi.it

Abstract. The paper presents the scenario of the Biomedical technologies in the frame of the Advanced Human-Machine Interface (HMI), with specific application to the direct Brain-Computer communication. This approach relies on the development of new miniaturized system for the unobtrusive measurement of biological signal using wearable or embedded sensors integrated in Advanced HMI to be used in performing a task. In the BCI application the actual goal is enabling the communication for severely disabled people with a future perspective to increase the possibilities offered by this technology and in rehabilitation and health care. This should be possible thanks to its integration in a more complex system of ambient intelligence allowing the control of primary functions at home or through the differentiation of specific system platform supporting other services.

7.1 Ambient Intelligence and Ubiquitous Computing Scenario

Ambient Intelligence (AmI) can be defined as a pervasive and unobtrusive intelligence in the surrounding environment supporting the activities and interactions of the users [1] in order to bring computation into the real, physical world and to allow people to interact with computational systems in the same way they would interact with other people.

The psychological and computational bases of this interaction touch aspects related both to the theory of mind and to the theory of computation. The subjects is not separated from the rest of the processes involved in the adaptation of the ambient functionalities. The Human Machine Interface (HMI) becomes a nested part of a cognitive system embedded in the ambient, and should tune itself to the user's needs.

The control of such a complex systems requires a highly embedded acquisition and computational infrastructure which implies the need of many

G. Andreoni et al.: *Human Machine Interface for Healthcare and Rehabilitation*, Studies in Computational Intelligence (SCI) **65**, 131–150 (2007)
www.springerlink.com

connections with the real world in order to participate in it. Of course there are different possible practical solutions, depending on the applications and on the users.

A general architecture requires a computation need to be everywhere in the environment, also embedded in clothes, to enable and facilitate people in the direct interaction with any kind of computational device; another model can be described only with sensors and transducers which are embedded in the environment, realizing a wireless or wired sensor network with a centralized processing and control device.

7.2 The Advanced Human Machine Interface Framework

HMI is a really generic term indicating a system which allows the creation of a low level man-machine interaction, like mouse and keyboards; advanced HMI (AHMI) are more complex systems that provide an high level of interaction moving towards a new definition of the interface concept. An important and really interesting topic is the introduction of several new items related to the optimization and the improvement of the communication process in order to support (Ambient Intelligence) applications. AHMI can allow the pursuit of an automated, adaptive system in order to guarantee the integration of human beings in the context surrounding them, to reach an **attuning** (mutual tuning) with the environment. Through attuning, the system is able both to induce, acquire and process emotional signals and to adapt its actions accordingly, with the ultimate goal of improving the effectiveness or the comfort of the supplied services. To exemplify this concept, we can consider a speech recognition system generally used to send commands or to write sentences; this system can also be enhanced to understand voice modifications in order to identify the frustration or fatigue during the different tasks. Other examples are related to imagines acquired during a videoconference, in which the computer can use the facial expression to make the communication with other people more efficient, by identifying their needs. It is worth noting that one of the most interesting aspects of the AHMI is the possibility to observe and identify the psycho physiological state of people improving the communication and the execution of the activities through an attuning between man and the machine. We maintain that such a high-level interaction has to be established through the identification of subjects condition with the important role played by the biosignals monitoring, in order to realize a real-time adaptation to human needs. New scenarios delineate a AHMI exploitation when human is aware that machine may understand him and cooperates with him in a task [2].

In this perspective, new interfaces with ambient intelligence have to behave like a skillful autonomous negotiation between the needs and the capabilities of the user. In particular it is conceivable to propose an interface providing the following distinguishing functionalities:

- two ways of interactivity;
- processing and interpretation of physiological and non verbal communication signals.

In this context it is really important to consider the interface configuration, comprising its input and output channels and which ones are available for the subject. This implies an Input/Output architecture which is strictly dependant on the context and on the applications: in a simple AmI context, movement and biosignals monitoring could provide enough information whilst an AmI scenario integrating an artificial cognitive system requires many different channels including voice recording, face expression and the biosignals analysis [4–6].

A generic architecture scheme in general could be composed by a controller, a feature extraction block, a data fusion block and an "intelligent" system that interprets the features and sends commands to the controller of the output channels, following the desired activity model. The user-interface primitives of these systems are not only menus, mice and windows but gesture, speech, affect, and context. The main outcome is the decision whether the subjects condition really requires the actuation of new strategies or the modification of some outputs, considering the provided models or the learned ones. This model assumes the use of Multi Modal Interfaces (MMI), which are defined as interfaces in which there is a concurrency of processing that enables the parallel utilization of the various communication channels; the presence of a data fusion block enables the combination of different types of data in order to facilitate and enhance input or output tasks [7, 8].

The massive use of input channels is requested in order to provide all the information needed by the algorithms to understand commands and realize the communication. MMI could be a powerful tool in AmI applications when a strong interaction with the subjects is required. Primary objective of MMI is to respond intelligently to speech and language, vision, gesture, haptics and other senses.

7.3 Brain Computer Interfaces (BCI)

7.3.1 Introduction to BCI Systems

A Brain Computer Interface (BCI) is a new human-machine communication system that doesn't rely on the normal output pathways of the central nervous system, like peripheral nerves and muscles, but introduces a direct communication channel between the brain and the external world [9]. The physiological activity of different areas of the brain are correlated either with the subject intention or with the interaction of the subject with the external world: for example the P300 is an event related potential which occurs when the subject attention is captured by a cognitive stimulus [10]. A BCI equipment exploits these phenomena trying to detect some modifications in the brain activity. Such patterns are defined *exogenous* when generated by specific external stimuli, *endogenous* if they are autonomically generated by the subject (during concentration or imagination tasks). The fundamental principle is to associate those variations with commands useful to interact with the external environment through a computer or similar devices.

Since '70, BCIs have been studied for military purposes. The poor performances in terms of information throughput and in terms of reliability didn't allow to reach the aim of driving and aircraft or activating a weapon. Scientists have explored the advantages and the applications of a direct connection between the brain and the external world in other fields, but nowadays the most important and fascinating application of a BCI is the aid to disabled people. Since there are many other technologies able to improve the residual abilities of the subject, the BCI address people with severe disabilities and whose residual capabilities are minimal.

Although many current research projects in the BCI field aim at the study of the interaction perspective to probe or explore brain functions, especially the dynamic mechanics and cooperation of different brain functional area, there are several potential applications related to the use of BCI as an advanced communication system.

From this point of view, the main fields addressed by BCI research, according to the current state of the art, could be summarized as follows:

Communication and Augmentative Environmental Control: dedicated to people with severe disabilities particularly, in order to increase or re-estabilish interaction with surrounding environments. This field comprises several methods of assistive communication, starting from simple binary capabilities and icons selection applications, up to virtual keyboards with spelling support;

Neural Prosthetics: in this particular case the output of a BCI system is used and optimized in order to provide control over an orthosis to those with mid-level cervical spinal cord injuries in order to restore the corrupted connection between the central nervous system and the affected limb or function;

Virtual Reality and Advanced Control Systems: mainly oriented at developing new control experience and innovative pheripherals in the gaming and amusement fields or in alternative clinical applications such as the biofeedback therapy.

More complex BCI applications targeted at a wider population of users and more complex tasks depend on the achievement of higher information transfer rates and better system stability. Thus accuracy, speed and reliability of this technology hinder the goals of its possible applications.

7.3.2 Structure of a Typical BCI System

As stated before a BCI have to exploit the information about the brain physiological activity in order to understand the user's intention. In order to perform such an operation, the system is typically composed by the following blocks (Fig. 7.1):

1. a data acquisition system;
2. a data processing real-time system;
3. a user interface.

The most diffused method for the acquisition of the brain activity is the electroencephalogram (EEG). The EEG signal reflects the electrophysiological activity of the brain and can be recorded by means of superficial electrodes applied on the scalp [11–13]. There are also other ways of recording the brain activity, but those techniques are invasive or require extremely expensive or bulky instrumentation.

The identification of the user's will from the signal relies on a data processing system: this system must process all the data as fast as possible in order to provide a real-time operation. The user have to perform some kind of activity in order to produce the signal associated to a specific command: each command is identified by a **class** of similar brain activities.

The first part is a block which performs a **data preprocessing** in order to reject artefact and to increase the signal to noise ratio (SNR), some kind of elaboration are band-pass filtering, adaptive filtering or envelope extraction.

After that a **feature extraction block** extracts the parameters (called 'features') from the signal which can be used to discriminate between different classes of possible commands. Typical features are the power density in certain area of the spectrum, the value and the index of the maximum and the root mean square amplitude.

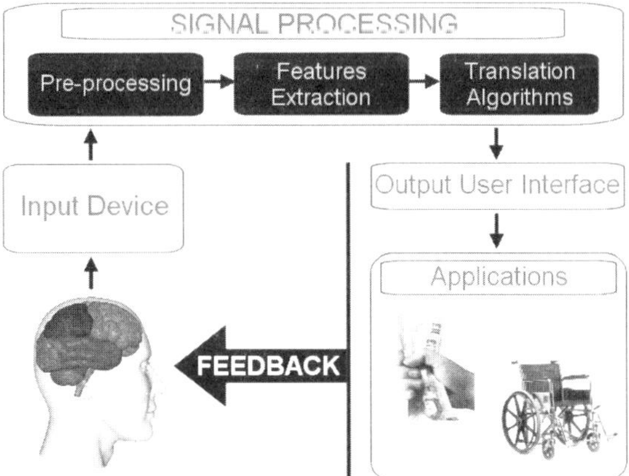

Fig. 7.1. Structure of a typical BCI system

The features vector identifies a multidimensional space inside which each class is situated in a different place. Given the coordinates of the signal in the space of the features, the **classifier** have to identify the discrimination boundaries between each area in order to be able to understand which class it corresponds to.

After the class is identified it is possible to associate it to a command. The command is executed by the user interface: it can be a word processing application, a functional electrical stimulation of some muscles or the activation of an home automation device. Usually a feedback device is used in order to allow the user to be informed about the quality of the produced signal and to understand the best mental strategy in order to produce a better signal in terms of separation between classes.

7.3.3 BCI Protocols

It is possible to identify two main approaches typically accepted in the specific BCI field, in order to generate those brain activity variations used as input control commands of a system. The first is the operant conditioning approach based on the self-driving and self-regulation of the EEG response. The second is the cognitive states detection approach (also known as pattern recognition approach) based on the identification of specific brain variations in response to a predefined cognitive mental task. Starting from this distinction, BCIs can then be categorized accordingly to what kind of imagery or mental tasks the users need to perform in order to drive or evoke the command-related EEG response. This choice is strictly correlated with the type of brain activity or neuromechanism that is going to be used as input in a BCI system. According to this categorization, the following typical BCI paradigms can be identified [9]:

- *mu rhythm control.* Similar to the alpha rhythm but with substantial physiological and topographic differences. It can be detected over the motor or somatosensory cortex in the 8-13 Hz frequency band. The subject can learn to control the amplitude of this waves in relation to actual and imagined limb movement or by performing intense mental tasks or just by increasing the subject selective attention.
- *Event related Synchronization/Desynchronization (ERS/ERD)* [14–16]. Those signal are recordable as movement related increments (ERS) or decrements (ERD) in specific frequency bands. They are localized over the sensorymotor cortex and their amplitude can be modified through an active feedback. Those signal exist even in the case of imagined movement. The cerebral activity related to the programming of a movement without the actual muscular activity is known as motor imagery, the preparation of the movement doesn't require any peripheral activity as it is just a preparation task. The main difference between imagined and executed movement is that the former is just a preliminary stage of the latter, whose evolution is blocked at a non precise cortico-spinal level.

- *P300* [9, 10]. Rare or particularly significant auditory, visual, or somato-sensorial stimuli, when interspersed with frequent or routine stimuli, typically evoked in the EEG over parietal cortex as a positive peak at about 300 ms. This wave is best recorded over the median line of the scalp.
- *Short latency Visual Evoked Potential (VEP)* [17]. VEP represent the exogenous response of the brain to fast and short visual stimuli which are recorded over the visual cortex corresponding to the occipital lobe of the scalp.
- *Slow Cortical Potential (SCP)* [9]. Among the lowest frequency features of the scalp recorded EEG are slow voltage changes generated in cortex consisting in potential shifts which occur over 300ms to many seconds.
- *Steady-State Visual Evoked Potentials (SSVEP)* [18]. Visually evoked potentials are the result of those electro-physiological phenomena that reflect visual information processing at a CNS level. The evoked response is considered transient if the stimulation frequency is less or equal to 2 Hz, steady-state if the visual stimulus repetition rate is greater than 6 Hz. In the latter case the resulting EEG signal shows an additional periodic response which leads to an increased spectral power localized on those frequencies integer multiples of the stimulus repetition rate. The elicited signal is particularly relevant over the occipital regions of the scalp. The amplitude of SSVEP is strictly related to user's attentive conditions thus it can be modulated and controlled by returning the subject an efficient feedback index.

7.3.4 Applications and Actual Performance of BCI Systems

Nowadays the consolidated outcomes of the worldwide research allow to conceive BCI as a potential actual aid for people at home. This fascinating perspective faces the problems related to those systems, which often are bulky and too expensive to become a practical solution for domestic appliance of BCI. Moreover the number of channel used by the algorithms should be minimized in order to simplify the application of electrodes; the feature extraction and classification algorithm should avoid redundancy in order to reduce the computational workload to the host computer. Nowadays some research groups have proposed a few studies carried out using miniaturized, battery powered EEG data acquisition systems. [19, 20]

In this chapter we can describe our experience, probably similar to those of other research teams in the world.

Example of 4-State PC-Driven BCI System Based on the SSVEP Protocol

The system was composed of an EEG helmet connected to a control software by a low-power RF connection. Considering the Human-machine Interface (HMI) as a control loop, the following hardware (HW) and software (SW) components have been designed and integrated in order to realize a multi-class wearable BCI system based on the SSVEP approach:

- a wearable device for EEG signal recording. The EEG signal is recorded by means of standard Ag/AgCl electrodes. The communication to the host PC is achieved by a RF Bluetooth® (BT) connection;
- a stimulation system composed of small sized light sources which can be easily applied to a standard LCD monitor;
- a three state protocol composed of a training, metatraining and validation phase;
- a BCI software composed of a C++ core which can control and communicate to the mathematical engine(MATLAB®, Mathworks Inc., Massachusetts, USA)for the offline analysis of the data acquired and for the single-sweep online classification of the signal.

The input device, called Kimera: operates a 12 bit signal conversion at 250sps through a miniaturized wireless transmission board, based on ARM® CPU and BT transmission. Two analog front-ends are devoted to the amplification and the filtering of the signal using a simple 3.3V supply voltage. The EEG signal are collected by means of standard Ag/AgCl electrodes applied using standard gel. According to the literature the preferred electrodes positions are based on the distribution of the visual pathways from eyes to the primary visual cortex. This implies a placement at O1 and O2 positions, according to the 10–20 International System of Electrode Placement.

The BCI software, Bellerophonte: It is characterized by multithreaded modular functions which makes it suitable for the implementation of many different protocols using the same HW acquisition system. A GUI module has been created in order to provide information, biofeedback and protocol management through a graphic user interface. A specific function has been developed in order to reduce the maximum desynchronization between the signal and the GUI generated triggers within four samples.

A thread is dedicated to the communication with the mathematical engine and to route the response of the algorithm to the appropriate module. A specific callback function is invoked by the Windows® multimedia timer and controls the user's stimulation circuitry.

The stimulation device: The visual stimulation system consisted of four cubic spotlights with sides of 2 cm, each cube can be attached to the four sides of a standard monitor, allowing the user to ideally associate each light to a simple direction: up, down, left or right (Fig. 7.2). Each light included a high efficiency green (wavelength 500 nm) led. In order to avoid direct exposition to the light and to diffuse the light beam in a more efficient way, we used a matt film to close the exposed face of the cube. The device accepted up to 8 LEDs and ensured the possibility to adjust their intensity.

The protocol: the system was based on a supervised translation algorithm and the protocol adopted consisted of three main functioning modes:

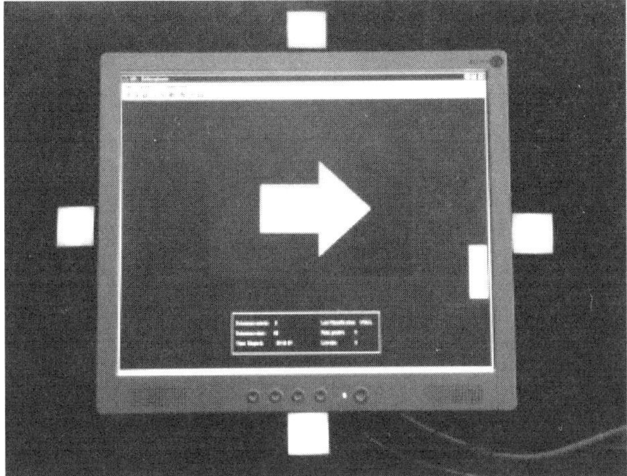

Fig. 7.2. The GUI and the stimulation unit during the meta-training phase

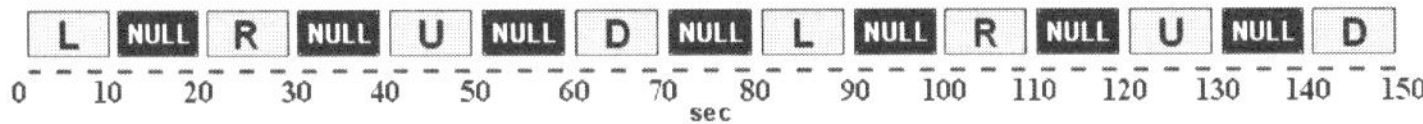

Fig. 7.3. Training session structures. The training session can be consistent or partitioned

- Training mode;
- Meta-training mode;
- Testing mode.

Training Mode: it acted as a data acquisition stage in order to record useful information for the later training of the supervised classifier. The interface guided the user through the training phase inviting him, by means of vocals instructions and a mobile cue at the centre of the screen, to gaze alternatively at the different light sources.

Two main alternative structures for the training session were provided: consistent and partitioned. Each structure scheme and the corresponding partial and total durations are summarized in Fig. 7.3.

The main advantages of a partitioned training structure consisted in the provision to the classifier of more information regarding the signal, during the onset and offset of each stimulus. Moreover it allowed the introduction of the possibility to train the classifier on data evoked by the same stimulus, but acquired during different time periods. It is worth noting that this particular training structure also led to less outwearing sessions, helping the user to maintain a stable concentration.

Meta-training Mode: in this configuration the system had been trained on the basis of the signal acquired during the previous stage, consequently it was ready for a real-time classification while the user is still guided by the BCI software. The user was asked to focus his attention on a particular light source while the signal was processed and identified continuously by the online translation algorithm; the switching from one stimulus to another occurred only after the system identified a command related to the actual target source. Together with the current estimated command (LEFT, RIGHT, UP, DOWN, NULL), the number of false positives and correct assignment were presented and continuously updated by the GUI.

Four level bars were placed in correspondence to each stimulus providing a biofeedback that acted as a quality and reliability index of the elicited signal identification.

False positive occurrence caused by the physiological time of reaction in stimulus fixation onset and offset and by the windowed data analysis was avoided introducing a latency between different classifications. In such period all the lights were switched off and the classification were not performed.

During this protocol stage the performances were evaluated on the basis of a system configured using only the training set. It was possible to use the acquired meta-training dataset in order to integrate the training data and to provide more information regarding hardly identified commands. Consequently this reduced information regarding these instructions that led to a fast identification as a confirmation of the training set reliability regarding the specific subset.

Testing Mode: the BCI system performs a continuous realtime classification of the signal, translating the estimated intention in a control command. In this configuration both stimuli related biofeedbacks and latency were active.

The classification algorithm: At the end of each training session the reliability of the acquired training set was evaluated by means of an offline analysis, aimed at obtaining a preliminary and quick assessment of the performances.

At the end of the offline analysis it was possible to specify which harmonics to consider during the training and the classification stages in order to maximize the reliability of the information.

The single-sweep signal processing and identification block included the following main steps (Fig. 7.4):

Signal pre-processing: the signal was high-pass filtered at 2Hz in order to avoid baseline fluctuations and to guarantee the typical SSVEP response band ($>6\,$Hz). It was also possible to activate an adaptive filtering (Adaptive Line Enhancer) with coefficients adjusted according to the LMS method. This

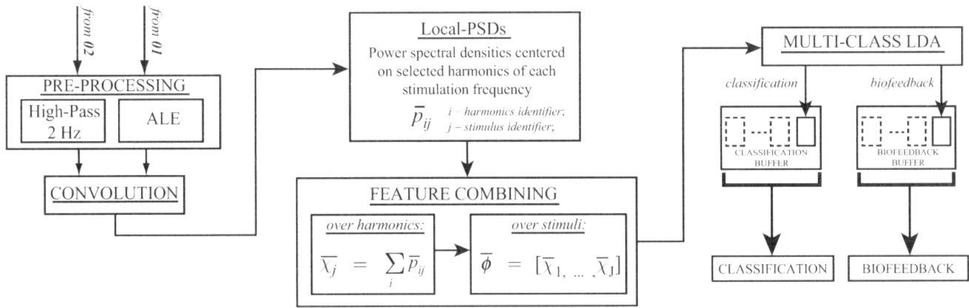

Fig. 7.4. Scheme of the signal processing, feature extraction and classification algorithm

allowed to highlight the typical SSVEP periodic components hidden in the basal activity and to obtain a better signal-to-noise ratio;

Feature extraction: the power spectral density of the convolution of the signal acquired from the two channels was calculated on a window of predefined size. The local-PSDs centered on the harmonics of interest of each stimulation frequency were combined point-to-point in order to limit the number of features for each stimulation frequency and to consider the information from multiple harmonics. The bandwidth of the considered local-PSDs was chosen accordingly to the effective frequency resolution and consequently to the minimum gap between the flashing frequencies in use and the analysis window size. This approach avoided confusing simple in-band amplitude variations (such as variations in the alpha band related, for example, to relaxation or attention) with the more localized (in frequency domain) steady-state visual evoked potential.

The classifier: it consisted of a regularized linear discriminant analysis (RLDA) based on the modified samples covariance matrix method. The RLDA included a boosting algorithm based on a cyclic minimization of the classification error on the training set and an algorithm for outliers rejection. The multi-class identification problem was solved by means of a combination of binary classifiers using a one-vs-all approach. The reliability in classification (feedback) was evaluated by combining the resultant time signed distance on all the boosting cycles with quality index related to the number of coherent identification obtained with the one-vs-all approach. The system was trained in order to identify five different classes referred to as LEFT, RIGHT, UP; DOWN and NULL for the nonstimulus class.

Postprocessing on classification outputs: two FIFO buffers of predefined length were continuously and respectively updated with the current feedback and classification values.

The effective classification and feedback values returned to the user were the results of the time-weighted combinations of the values contained in each buffer. In this way it was possible to analyze a specific portion in the past of the recent classification values in order to enhance the system stability in

142   G. Andreoni et al.

terms of false positives thus allowing a smoother perception of the achieved control. In case of uncertainty, the system forced the classification to NULL.

Results: the system showed good robustness against false positives and was able to achieve an accuracy (calculated over the meta-training data-set) between the 80% and 100%. The average speed was of 10–15 commands per minute, depending on the response of the subject to the visual stimulation in terms of number of harmonics and amplitude of the signal. The performance analysis was based on the system control level perceived by the user: the NULL classification was considered as an idling condition so it was not involved in the characterization of the system in terms of accuracy, but it was indirectly considered in terms of bit-rate. The considered features revealed to be meaningful for the specific application: the Fig. 7.5 shows the mean values of the features during four different frequency stimulation (6, 7, 8 and 10Hz) compared to the idling phase.

It was also possible to verify the reliability of the stimulation with computers of different price range: it is worth underlying the importance of a controlled environment in terms of brightness: it isn't necessary to work in a dark room, but a softly enlightened environment ensure and higher contrast of the light spots.

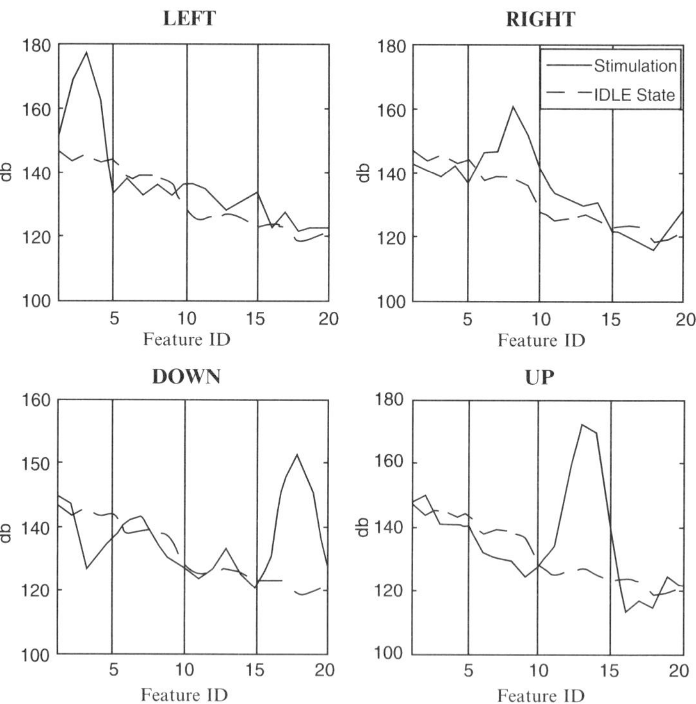

Fig. 7.5. Average values of the features during four different frequency stimulation (6, 7, 8 and 10Hz) referred to the idling state

The implemented SSVEP protocol lead to an easy to use communication system which required only the application of four electrodes: the four commands based communication provided an increase in communication speed making it possible to control more complex interfaces. The user-friendliness and the low cost of the proposed platform will make it suitable for the development of home BCI applications and long term application studies.

7.3.5 The Next-Gen Advanced HMI

The future will be the complete development of an intelligent system able to allow any people, also affected by severe motor disabilities, to drive a wide set of electro-mechanical devices.

This particular research field aims at implementing the complete "From Intention to Action" chain (Fig. 7.6) allowing any people to manage and use his/her own control system, taking in account his/her own available resources (i.e. possible control signals) and his/her own specific needs (i.e. most frequently performed actions, priority of tasks to execute, and so on).

The interaction between the user and the system will be so complex that the user will no longer be obliged to adhere to strict formal communication protocols imposed by the machine. The system must be able to support some kind of intelligent 'declarative' interaction. In such a context the machine must be capable of understanding the intentions of the user.

The most important gap in our present understanding of the process in human-human communication is the lack of a comprehensive theory of the ways in which the information coming from the five senses (auditory, visual, tactile) is integrated with the enormous archive of prior knowledge related to interactions and the world in general.

It is difficult to imagine how a coherent perception of intentions can be synthesized without such an integration or fusion. The single input from individual channels may be highly ambiguous. Straightforward, linear combination of all inputs does not guarantee disambiguation; in some cases the user

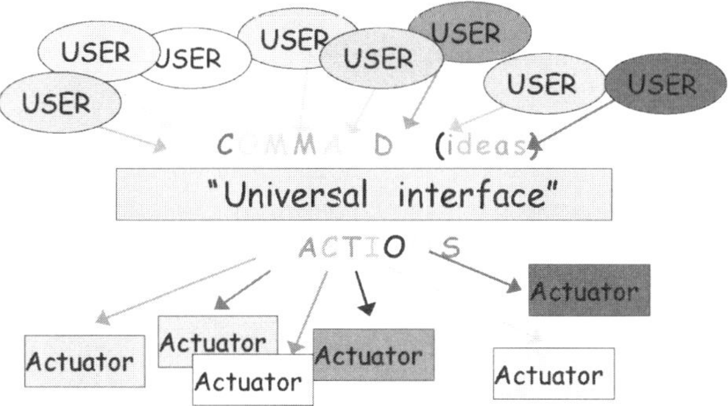

Fig. 7.6. The vision of the "from Intention to Action" paradigm

simply does not provide all the information that is necessary to understand the intention, in other cases the different channels may seem to carry contradictory contributions.

For all that is currently known about the ways in which humans understand intentions, it is clear that understanding is an active process (involving both signal-driven and interpretation-driven processes). Active processes are part and parcel of all individual components, even if they have been largely disregarded or modelled very poorly, like in existing speech recognition and natural language processing.

To get these ambitious targets, different areas of applied technologies should be approached:

1. User interfaces design: wearable sensors and microelectronics technologies should supply the way to collect and interpret any biological signal that could allow the user to "communicate" an intention. It's a basic step towards the collection of as many signals as possible: the availability of a wide set of distinguishable commands could speed up the interfacing with the drivable devices;

2. Human Machine Interface design: the core of the system, i.e. the "universal interface" is an intelligent unit able to recognize any user command and perform it. Such a unit should be able to recognise its "master" and, accordingly to his preferences and needs, to perform different interpretations of the sequence of input signals. Prediction and learning techniques could be used to design a software interface always conscious of its "master" requests. The shortest path from idea to action should be always implemented; Harmonization and standardization of the communication protocols towards "end actuators" to perform the requested action (i.e. home appliances, domotic devices, electronic wheel chairs, personal communicators...) to make the system really usable and easily spreadable all over the world.

A Future perspective for Next-Gen BCI: fNIRS-EEG Multi-Modal BCI

In the fundamental research about subject's intention detection and communication recovery, we have seen that most of the actual generation of BCI system are EEG-based: this means that the base technology used is electroencephalography or Electro-corticography, but several criticism in these systems still exist both on the hardware side and on the software one. The main problems in actual BCI systems are:

1) cost of device (today they require an 'open' EEGraph);
2) the possibility to measure and classify in real time specific cortical events;
3) the low communication rate;
4) the accuracy of the classification.

For example noninvasive functional human brain mapping by diffuse optical methods is a novel technique that employs near infrared light to probe the brain for changes in parameters relating to brain activity and it could be applied to re-enforce the EEG-BCI to obtain a more reliable and efficient BCI. The multimodal detection of the subject's intention could lead to a new understanding of the brain physiology but also to the realisation of a next-gen BCI platform.

One possible direction will be the design and the integration of BCI EEG-Based and fNIRS-based in one portable devices to offer an absolutely innovative opto-electric-functional device for brain activity investigation and to explore the possibility to increase BCI accuracy and communication speed through the integration of several cerebral signals. This will not solve neither the expensiveness of the hardware nor the usability (easy-to-use issue) of the system but could significantly improve the efficiency of such systems.

Starting almost 30 years ago with the pioneering work of Jobsis [21] noninvasive near-infrared spectroscopy (NIRS) has been used first to investigate experimentally and clinically brain oxygenation in neonates and adults, and later to assess muscle oxidative metabolism in pathophysiology.

NIRS employs optical radiation in the range 600–1000 nm where light attenuation by tissue constituents (water, lipid, oxyhemoglobin and deoxyhemoglobin) is relatively low and accounts for an optical penetration through several centimeters of tissue. The difference in the absorption spectra of oxyhemoglobin (O2Hb) and deoxyhemoglobin (HHb) allows the separate measurements of the concentration of this two species, and the derivation of physiologically relevant parameters like total hemoglobin content (tHb = HHb + O2Hb) and blood oxygen saturation (SO2 = O2Hb/tHb).

So far, the most common approach of NIRS is in the continuous wave (CW) regime, that is to measure light attenuation at two wavelengths between a couple of optical fibres set normal to the tissue surface at a known relative distance (typically 2–4 cm), to assume a fixed value of the scattering coefficient (in the spectral range 600–1000 nm biological tissues are in fact highly diffusive media), and, applying the Lambert-Beer law, to track the relative changes in HHb and O2Hb. This rather straightforward technique has grown in time and led to instruments working at more wavelengths, or with multiple sources, and multiple detectors geometries. In particular, this latter feature offers the possibility to probe simultaneously different regions in the tissue under study and thus to produce functional maps.

In spite of the most powerful high cost tools for human brain mapping such as positron emission tomography (PET) and functional magnetic resonance imaging (fMRI), since 1993 functional NIRS (fNIRS) using continuous wave (CW) has been largely used to investigate the functional activation of the human cerebral cortex [22]. fNIRS allows the investigation (without discomfort and interference related to the intrinsic limitations of PET and fMRI) of regional concentration changes in O2Hb and HHb mainly in small vessels, such as the capillary, arteriolar and venular bed of the brain cortex.

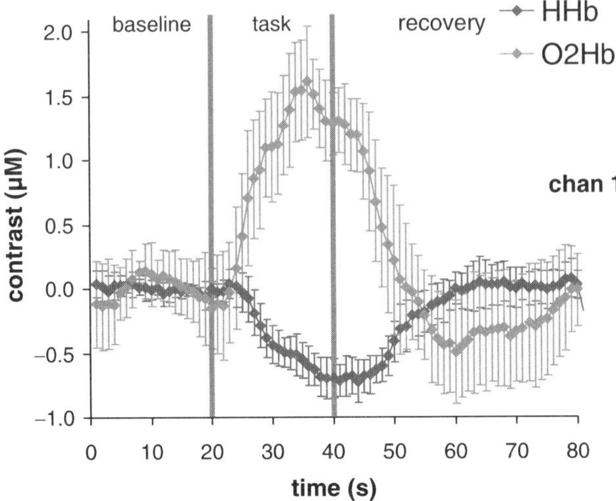

Fig. 7.7. Time course of ΔO2Hb (open symbols) and ΔHHb (filled symbols) during the exercise performed with the right hand (triangle) and with the left hand (square). The vertical lines represent the task interval. Each curve is an average of the five measures performed with the same hand. A mobile average of 4 s was applied

The physiological meaning of O2Hb increase and HHb decrease upon cortical activation (Fig. 7.7) has recently been extensively investigated in a rat brain model [23] and in humans [24]: O2Hb is the most sensitive indicator of the changes in cerebral blood flow and the direction of the changes in HHb is determined by the degree in venous blood oxygenation and volume. In addition, studies combining either PET or fMRI with fNIRS demonstrated that the oxygenation changes measured by fNIRS correspond to signal intensity increases detected by fMRI and PET [25].

In the last years some multi-channel CW fNIRS systems have been developed and tested to provide spatial maps (with a spatial resolution of about 0.5 cm) of oxygenation changes of frontal, temporal, parietal, and visual cortical areas upon different stimuli [26]. The key limitations of CW fNIRS technique are the coupling between the absorption and the scattering coefficient causing the lack of quantitative assessment, sensibility to artefacts and experimental conditions. A possible way to uncouple absorption from scattering is based on the use of sinusoidally modulated laser sources, and on the measurement of signal phase and amplitude changes caused by propagation. The so called frequency-domain (FD) technique extracts both the mean scattering coefficient and absorption coefficient of the probed medium provided that a non-trivial calibration of collection efficiencies is performed. This approach was successfully applied for measuring tissue oxygenation, and led to the first identification of a rapid change of scattering coefficient following a cerebral stimulus [27]. The dual approach is the study of photon migration in the time domain, first introduced in vivo by Chance et al [28] and

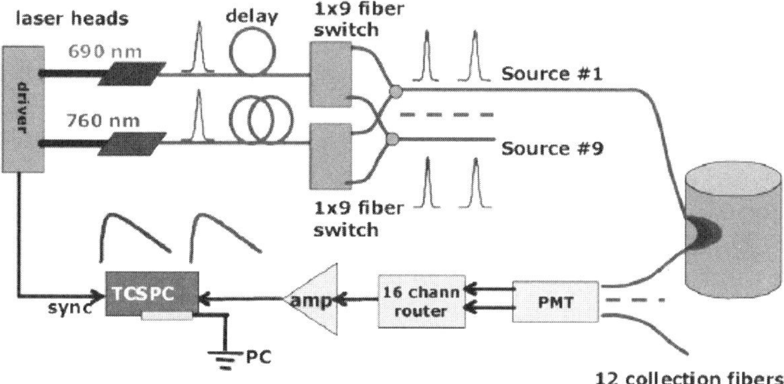

Fig. 7.8. Scheme of a nine-source and 12-detector TR tissue oximeter. TCSPC: time-correlated single photon counting; PMT: photomultiplier tube

Delpy et al [29]. This approach is based on the detection of attenuation, broadening and delay experienced by a short (few thousands of picoseconds) laser pulse injected in a diffusive medium. In the past, the cost, size, and complexity of time-resolved (TR) instrumentation have prevented this technique from being effectively used in tissue oxygenation studies so far. Some authors recently reported on the development of a compact multi-channel time-resolved tissue oximeter. The system operates with 2 wavelengths, 2 injection points and 8 independent collection points, and a typical acquisition time of 166 ms (Fig. 7.8).

They have successfully used this system for mapping of calf muscle during plantar flexion exercise [30] and for measuring brain activity associated with letter fluency [31]. Recent studies have foreseen and demonstrated that TR fNIRS system may have other advantages over CW fNIRS systems especially in terms of increased penetration depth. In particular, differently from CW systems, where sensitivity to depth is obtained by increasing the distance between source and detector, in TR system sensitivity to depth can be obtained at a fixed source detector distance by selecting late photons.

In the case of human brain mapping, since the human head has a layered structure (roughly skin, skull and brain), a better discrimination of the signal from the deep layer of brain cortex from the noise due to the systemic activity in the more superficial layers of the skin can be achieved.

While in recent years fNIRS studies have been steadily growing in the biomedical community, at both the research and clinical level, the use of fNIRS for BCI studies is in a very preliminary stage. To our knowledge in fact only a case study has been recently published [32] in which a single channel CW NIRS system has been used to monitor hemodynamic response following motion imagery (i.e. the imagination of physical movement) on volounters. The potential of fNIRS for BCI studies is therefore almost unexplored, not only as a stand alone technique but also in combination with other modalities like EEG and fMRI.

7.4 Conclusions

This chapter wants to address the recent developments and research trends in Advanced Human Machine Interaction with specific reference to rehabilitation and healthcare. Probably the most exciting goals of the research are represented by the possibility to establish a direct communication between a machine and the human brain so that to overcome possible interruption of the "traditional" communication pathways and allowing new ways of interaction.

In this meaning, a Brain Computer interface can be a solution for the communication of people with severe disabilities, messages and commands can be expressed by electrophysiological phenomena that are related to the user's intention. The state of the art BCI systems provide a slow and poorly accurate communication channel, but the research in terms of algorithms and electronic equipment are focused on solving this kind of problems. In short term, it doesn't seem reasonable to imagine a BCI to drive a car or as a substitute to the traditional pathways of the human body, however it can improve the quality of life to people whose interaction with the external environment is severely affected.

Acknowledgments

This work was partially supported by a grant from IIT.

References

1. Riva, G., Loreti, P., Lunghi, M., Vatalaro, F., Davide, F.(eds): Presence 2010: The Emergence of Ambient Intelligence. IOS PRESS, Amsterdam (2003)
2. Greene, J.O.: Message production. Advances in Communication Theory. Erlbaum, Mahwah N.J. (1997)
3. Andreoni, G., Anisetti, M., Apolloni, B., Bellandi, V., Balzarotti, S., Beverina, F., Campadelli, P., Ciceri, M. R., Colombo, P., Fumagalli, F., Palmas, G., Piccini, L.: Emotional interfaces with ambient intelligence (in press on: Athanasios Vasilakos and Witold Pedrycz (eds), Ambient Intelligence, Wireless Networking, and Ubiquitous Computing, 2006)
4. Picard, R., Vyzas, E., Healey, J., Toward machine emotional intelligence: analysis of affective physiological state, IEEE Trans. Pattern Anal. Mach. Intell., vol. 23, no. 10, (2001) 1175–1191
5. Murphy, R, et al., Emotion-based control of cooperating heterogeneous robots, IEEE Transactions on Robotics and Automation, vol. 18, no. 5, (2002) 744–757
6. Nasoz F., Alvarez K., Lisetti C.L., Emotion recognition from physiological signals for presence technologies, International Journal of Cognition, Technology, and Work – Special Issue on Presence, vol. 6, no. 1, 2003
7. Coutaz, J., and Nigay, L., A design space for multimodal systems concurrent processing and data fusion, in INTERCHI'93 Human Factors in Computing Systems, 1993, pp. 172§–178

8. Gaver W., Synthesising auditory icons, in INTERCHI'93 Human Factors in Computing Systems, 1993, pp. 228§–235

9. Wolpaw, J. R., Birbaumer, N., McFarland, D.J., Pfurtsheller, G., Vaughan, T.M., Brain Computer Interfaces for communincation and control, Clinical Neurophysiology, 113:767–791, 2002

10. Beverina, F., Silvoni, S., Palmas, G., Piccione, F., Giorni, F., Tonin, P., Andreoni G., P300-based BCI: a real-time working environment to test HCI on healthy and tetraplegic subjects. Biomedizinische Technik, Band 49, Ergänzungsband 1, 2004 pp. 35–36

11. Kennedy P., Bakay R., Restoration of neural output from a paralyzed patient by a direct brain connection. NeuroReport, Vol. 9: (1998); 1707–1711

12. Kennedy P., Bakay R., Moore M., Adams K., Goldwaithe J., Direct control of a computer from the human central nervous system. IEEE Trans Rehabil Eng 2000;8:198–202

13. Wolpaw J., McFarland D., Neat G., Forneris C., An EEG-based brain– computer interface for cursor control. Electroenceph clin Neurophysiol 1991;78:252–259

14. Pfurtscheller G., Lopes da Silva F., Event-related EEG/MEG synchronization and desynchronization: basic principles. Clin Neurophysiol 1999;110:1842–1857

15. Pfurtscheller G., Neuper C., Schloegl A., Lugger K., Separability of EEG signals recorded during right and left motor imagery using adaptive autoregressive parameters. IEEE Trans Rehabil Eng 1998;6:316–325

16. Pfurtscheller G., Neuper C., Motor imagery and direct brain-computer communication. Proc IEEE 2001;89:1123–1134

17. Sutter E., The brain response interface: communication through visually induced electrical brain responses. J Microcomput Appl 1992;15:31–45

18. Andreoni G., Beverina F.D.B., Palmas G., Silvoni S., Ventura G., Piccione F., BCI based on SSVEP: methodological basis. Biomedizinische Technik, Band 49, Ergänzungsband 1, 2004 pp. 33–34

19. Maggi L., Piccini L., Parini S., Beverina F.D.B., Silvoni S., Andreoni G., A portable electroencephalogram acquisition system dedicated to the Brain Computer Interface. Biomedizinische Technik, Band 49, Ergänzungsband 1, 2004 pp. 69–70

20. Guger, C., Edlinger, G., Krausz, G., Laundl, F., Niedermayer, I., Architectures of a PC and Pocket PC based BCI system, Proceedgins of the 2nd International Brain-Computer Interface Workshop and Training Course, pp. 49–50, September 2004

21. Jobsis F.F., VanderVliet, "Discovery of the near-infrared window into the body and the early development of near-infrared spectroscopy", JBiomedOpt.4: 392–396 (1999)

22. Villringer A., et al., "Near infrared spectroscopy (NIRS): a new tool to study hemodynamic changes during activation of brain function in human adults", NeurosciLett 154:101–104 (1993)

23. Hoshi Y., et al., "Interpretation of near-infrared spectroscopy signals: a study with a newly developed perfused rat brain model", JApplPhysiol 90:1657–1662 (2001)

24. Boas D.A., et al., "The accuracy of near infrared spectroscopy and imaging during focal changes in cerebral hemodynamics", Neuroimage 13:76–90 (2001)

25. Rostrup E., et al., "Cerebral hemodynamics measured with simultaneous PET and near-infrared spectroscopy in humans", BrainRes 954:183–193 (2002)

26. Jasdzewski G., et al., "Differences in the hemodynamic response to event-related motor and visual paradigms as measured by near-infrared spectroscopy", Neuroimage 20:479–488 (2003)
27. Gratton E., et al., "Measurements of scattering and absorption changes in muscle and brain", PhilTransRSocLondonBiolSci 352:727–735 (1997)
28. Chance B., et al., "Comparison of time-resolved and -unresolved measurements of deoxyhemoglobin in brain", ProcNatlAcadSciUSA 85:4971–4975 (1988)
29. Delpy D.T., et al., "Estimation of optical pathlength through tissue from direct time of flight measurement", PhysMedBiol 33:1433–1442 (1988)
30. Torricelli A., et al., "Mapping of calf muscle oxygenation and haemoglobin content during dynamic plantar flexion exercise by multi-channel time-resolved near infrared spectroscopy", PhysMedBiol 49:685–699 (2004)
31. Quaresima V., et al., "Bilateral prefrontal cortex oxygenation responses to a verbal fluency task: a multi-channel time-resolved near-infrared topography study", JBiomedOpt 10:011012 (2005)
32. Coyle S., et al., "On the suitability of near infrared (NIR) systems for next generation brain computer interfaces", PhysiolMeas 25:815–822 (2004)